Ionic Conductive Polymers for Electrochemical Devices

Ionic Conductive Polymers for Electrochemical Devices

Editor

Riccardo Narducci

MDPI • Basel • Beijing • Wuhan • Barcelona • Belgrade • Manchester • Tokyo • Cluj • Tianjin

Editor
Riccardo Narducci
Industrial Engineering
University of Rome "Tor
Vergata"
Roma
Italy

Editorial Office
MDPI
St. Alban-Anlage 66
4052 Basel, Switzerland

This is a reprint of articles from the Special Issue published online in the open access journal *Polymers* (ISSN 2073-4360) (available at: www.mdpi.com/journal/polymers/special_issues/Polym_Electrochem).

For citation purposes, cite each article independently as indicated on the article page online and as indicated below:

LastName, A.A.; LastName, B.B.; LastName, C.C. Article Title. *Journal Name* **Year**, *Volume Number*, Page Range.

ISBN 978-3-0365-2941-7 (Hbk)
ISBN 978-3-0365-2940-0 (PDF)

Contents

About the Editor

Riccardo Narducci

Riccardo Narducci received his Master's degree in Chemistry at the University of Perugia (2004). He obtained a double Ph.D. in Chemistry (Cum Laude) at the University of Rome "Tor Vergata"and Aix-Marseille Université. He was H-Chercheur at Aix-Marseille Université and adjunct Professor at the University of Tor Vergata from 2011 to 2014. He won the Vinci Programme (UFI) for 2013 and 2016. He has participated in five European Programs on PEMFCs. He is a Review Editor of *Frontiers in Energy Research* (EPFL), and Guest Editor of *Membranes* (Topic Editor), *Polymers* and *Crystals* (MDPI). He is an author and co-author of about 50 research articles, reviews, and book chapters on international journals, one Italian patent, one international patent-pending, and more than 45 scientific communications (keynote, invitations, etc.) in international congress and meetings. Currently, he is a member of LIME and is involved in synthesis and characterization of anionic and ampholytic membranes for FCs, inorganic materials, and development of INCA method for PFSA membranes. From January 2020, he obtained habilitation as Associate Professor in Chemistry (SC 03/B2, SSD Chim07).

Editorial

Ionic Conductive Polymers for Electrochemical Devices

Riccardo Narducci [1,2]

[1] Department of Industrial Engineering, University of Rome "Tor Vergata", 00133 Rome, Italy; riccardo.narducci@uniroma2.it

[2] International Laboratory Ionomer Materials for Energy (LIME), 00133 Rome, Italy

Citation: Narducci, R. Ionic Conductive Polymers for Electrochemical Devices. *Polymers* **2022**, *14*, 246. https://doi.org/10.3390/polym14020246

Received: 22 November 2021
Accepted: 21 December 2021
Published: 7 January 2022

Increasing levels of pollution (especially in large cities), the rising cost of oil, and climate change are pushing the scientific community towards more sustainable solutions for the conversion and storage of energy. Devices such as fuel cells (FCs), redox flow batteries (RFBs), and electrolyzers can help to significantly decrease the amount of greenhouse gases emitted. Ionic conductive polymers are fundamental components of these devices (protonic, anionic, and amphoteric), generally requiring great chemical and mechanical stability; good performance and durability; low permeability to reagents; and excellent characteristics of weight, volume, and current density for several applications from mobile to automotive and co-generation systems. Unfortunately, the high cost of perfluorinated ionomers and the low stability of anionic polymers in alkaline environment, among other things, still limit their use. This Special Issue of *Polymers* is dedicated to this exciting research field, with some excursions in related fields, focusing on commercial polymers such as Nafion, a benchmark for proton conducting membranes, acid doped polybenzimidazole (PBI), or blended membranes containing hyperbranched PAES/Linear PPO as anion exchange membranes (AEMs). Promising and low-cost sulfonated aromatic polymers (SAP), such as sulfonated poly(ether ether ketone) (SPEEK) and sulfonated poly(phenyl sulfone) (SPPSU) ionomers in phosphate buffer solution for enzymatic fuel cell, or crosslinked sulfonated polyphenylsulfone-vinylon (CSPPSU-vinylon) are presented. The ecofriendly poly(vinyl alcohol) (PVA) used to modify the catalyst layer, or the use of novel ionic liquid-incorporated Zn-ion conducting polymer electrolyte membranes, or solid polymer blend electrolytes (SPBEs) based on natural chitosan (CS) and methylcellulose (MC) are proposed. In this book, we also report some strategies to enhance the mechanical stability, such as cross linking (XL), or several techniques, including classical casting methods or electrospinning (ES), used to obtain ionomer nanofibers with different morphologies depending on relative humidity (RH); all properties were studied with classical investigations techniques, such as impedance, dynamic mechanical analysis (DMA), FC tests, or the new INCA method. To reflect the broad scope of this topic, contribution from leading scientist across the world, whose research addresses ionomeric membranes and catalysts from different perspectives, sharing a common vision of pollution reduction and the search for sustainable energy sources, have been gathered. In this manner, Lufrano et al. [1] investigated how Nafion 1100 membrane preparation procedures affect both the morphology of the polymeric film and the proton transport properties of the electrolyte. A comparison between commercial membranes such as Nafion 117 and Nafion 212 and Nafion membranes prepared by three different procedures, Nafion-recast, Nafion uncrystallized, and Nafion 117-oriented, was conducted. The conductivity measurements show that the anisotropy increased from ~20% in the commercial membrane up to 106% in the pressed membrane (Nafion 117-oriented) where the ionic clusters were averagely oriented parallel to the surface, leading to a strong directionality in proton transport. The solution cast Nafion uncrystallized membrane showed the lowest water diffusion coefficients, conductivities, and Young's modulus, highlighting the correlation between low crystallinity and a more branched and tortuous structure of hydrophilic channels. The authors suggested to avoid the use of both materials, oriented and uncrystallized, in high-temperature fuel cells. Pasquini et al. [2] studied the hydrolytic stability, conductivity, and mechanical behavior of SPEEK and SPPSU ionomers in phosphate buffer solution for enzymatic fuel cell. The results showed that the membrane stability can be adapted by changing the casting solvent (water, ethanol or DMSO) and procedures. A shorter casting time resulted in stiffer membranes but had no effect on the mass uptake (MU) and conductivity. The crosslinking stabilized the membranes that showed a better hydrolytic stability, even if with a little decrease of conductivity. The addition of SPPSU to SPEEK membranes improved the ionic conductivity. Kim et al. [3] prepared a series of novel AEMs with hyperbranched brominated poly(arylene ether sulfone) (Br-HB-PAES) and linear chloromethylated poly(phenylene oxide) (CM-PPO) with different weight ratios followed by quaternization with triethylamine, which promoted the ion channel formation. In particular, the Q-PAES/PPO-55 membrane showed a very high hydroxide ion conductivity (0.91×10^{-1} S cm^{-1}) around three times higher than the pristine Q-HB-PAES membrane. In addition, the rigid hyperbranched structure showed an enhancement of the swelling ratio and demonstrated an alkaline stability under 2M KOH conditions over 1000h at 50 °C. Jienkulsawad et al. [4] employed, in the fabrication of membrane electrode assemblies (MEAs), a PVA to humidify the membrane in proton-exchange membrane fuel cells (PEMFCs) operated under low-humidity conditions. The 0.03 wt% PVA in the anode catalyst layer (CL) and 0.1 wt% PVA on the gas diffusion layer (GDL) improve the current density by approximately 30% at the operating voltage of 0.6 V and non-humidified anode and cathode humidifier temperature of 25 °C. Liu et al. [5] prepared a novel membrane containing a polymer matrix poly(vinylidene fluoride-hexafluoropropylene) (PVdF-HFP) and 1-ethyl-3-methylimidazolium trifluoromethanesulfonate (EMITf), along with zinc trifluoromethanesulfonate $Zn(Tf)_2$. The best amorphous and nanopored membrane, ILPE-Zn-4, with a mass ratio of 0.4:0.4:1 (EMITf:$Zn(Tf)_2$:PVDF-HF), showed at room-temperature (RT) ionic conductivity of ~1.44×10^{-4} S cm^{-1} with a wide electrochemical stability window (~4.14 V) and thermal decomposition temperature ~305 °C with good tensile strength ~5.7 MPa. This polymer electrolyte could be a promising candidate for energy storage applications. Aziz et al. [6] studied and synthesized, with a solution cast technique, SPBEs materials based on CS and MC incorporated with different concentrations of ammonium fluoride (NH_4F) salt. The electrochemical stability of the electrolyte sample was found to be up to 2.3 V via the linear sweep voltammetry (LSV) study. The value of specific capacitance was determined to be around 58.3 F g^{-1}. The synthesized electrical double-layer capacitor (EDLC) cell was found to exhibit high efficiency (90%). In the first cycle, the values of internal resistance, energy density, and power density of the EDLC cell were determined to be 65 Ω, 9.3 Wh kg^{-1}, and 1282 W kg^{-1}, respectively. Escorihuela et al. [7] presented a systematic study of the physicochemical properties and proton conductivity of PBI membranes doped with common phosphoric acid at different concentrations, 0.1, 1, and 14M, and with other alternative acids such as natural phytic acid (0.075M) and phosphotungstic acid (HPW, 0.1M). The cross-section SEM images show the formation of channels in the polymeric network thanks to the use of these acids, also keeping their mechanical properties and thermal stability. Under low acid doping (0.1M), membranes doped with phytic acid displayed a superior conducting behavior (2.6×10^{-4} S cm^{-1}), compared to doping with phosphoric acid (5.8×10^{-6} S cm^{-1}), proved to be a sustainable alternative. Kim et al. [8] synthesized a thermally crosslinked sulfonated polyphenylsulfone (CSPPSU) polymer and different wt% of PVA (5, 10, and 20); then, a CSPPSU-vinylon membrane was synthesized using a formalization reaction. The conductivity of the CSPPSU-10vinylon membrane reached 0.66×10^{-1} S cm^{-1} at 120 °C under 90% RH and was higher than CSPPSU membrane. The fuel cell results showed higher current densities than those of Nafion 212 and CSPPSU membranes, obtained under high- and low-humidification conditions. These results are due to the excellent water retention even under low humidification conditions of vinylon membranes and can be proposed as an alternative to fluoropolymer electrolytes, especially for thin membranes. Halabi et al. (Dekel's group) [9] prepared, by ES technique, anion-conducting ionomer-based nanofibers with different morphologies depending on the RH during the process (RH_{ES}). The formation of branched thin fibers was observed in mats prepared at RH_{ES} 20% and 30%. This affects the water uptake (WU) and conductivity, which are higher for fibers formed at low humidity, attributable to their larger diameter. The understanding of these parameters is important for the future design of these devices with tailored properties. Raja Rafidah et al. [10] discussed the recent progress of aromatic-based membranes that represent some of the best alternatives in proton exchange membranes (PEMs) due to their electrochemical, mechanical, and thermal strengths. Membranes based on these polymers, such as poly(aryl ether ketones) (PAEKs) and polyimides (PIs), however, lack a sufficient level of proton conductivity and durability. Various strategies are proposed to improve these characteristics: crosslinking, multiblock copolymerization, the introduction of inorganic/organic fillers/nanofillers, etc. Although they have disadvantages that limit their use in fuel cells, aromatic-based polymers still hold great potential as effective and low-cost alternatives to perfluorinated ones.

I am confident that the articles contained in this Special Issue will serve to further stimulate advances in this research area, in both the sectors of membranes and catalysts; the first is essential for the long-term functioning of the system, and the second for a drastic reduction in costs, especially in fuel cells. I thank all my friends and colleagues who contributed papers to this Special Issue.

Funding: This research received no external funding.

Conflicts of Interest: The author declares no conflict of interest.

References

1. Lufrano, E.; Simari, C.; Di Vona, M.L.; Nicotera, I.; Narducci, R. How the Morphology of Nafion-Based Membranes Affects Proton Transport. *Polymers* **2021**, *13*, 359. [CrossRef] [PubMed]
2. Pasquini, L.; Zhakisheva, B.; Sgreccia, E.; Narducci, R.; Di Vona, M.L.; Knauth, P. Stability of Proton Exchange Membranes in Phosphate Buffer for Enzymatic Fuel Cell Application: Hydration, Conductivity and Mechanical Properties. *Polymers* **2021**, *13*, 475. [CrossRef] [PubMed]
3. Kim, S.H.; Lee, K.H.; Chu, J.Y.; Kim, A.R.; Yoo, D.J. Enhanced Hydroxide Conductivity and Dimensional Stability with Blended Membranes Containing Hyperbranched PAES/Linear PPO as Anion Exchange Membranes. *Polymers* **2020**, *12*, 3011. [CrossRef] [PubMed]
4. Jienkulsawad, P.; Chen, Y.-S.; Arpornwichanop, A. Modifying the Catalyst Layer Using Polyvinyl Alcohol for the Performance Improvement of Proton Exchange Membrane Fuel Cells under Low Humidity Operations. *Polymers* **2020**, *12*, 1865. [CrossRef] [PubMed]
5. Liu, J.; Ahmed, S.; Khanam, Z.; Wang, T.; Song, S. Ionic Liquid-Incorporated Zn-Ion Conducting Polymer Electrolyte Membranes. *Polymers* **2020**, *12*, 1755. [CrossRef] [PubMed]
6. Aziz, S.B.; Hamsan, M.H.; Nofal, M.M.; San, S.; Abdulwahid, R.T.; Raza Saeed, S.; Brza, M.A.; Kadir, M.F.Z.; Mohammed, S.J.; Al-Zangana, S. From Cellulose, Shrimp and Crab Shells to Energy Storage EDLC Cells: The Study of Structural and Electrochemical Properties of Proton Conducting Chitosan-Based Biopolymer Blend Electrolytes. *Polymers* **2020**, *12*, 1526. [CrossRef] [PubMed]
7. Escorihuela, J.; García-Bernabé, A.; Compañ, V. A Deep Insight into Different Acidic Additives as Doping Agents for Enhancing Proton Conductivity on Polybenzimidazole Membranes. *Polymers* **2020**, *12*, 1374. [CrossRef] [PubMed]
8. Kim, J.-D.; Matsushita, S.; Tamura, K. Crosslinked Sulfonated Polyphenylsulfone-Vinylon (CSPPSU-vinylon) Membranes for PEM Fuel Cells from SPPSU and Polyvinyl Alcohol (PVA). *Polymers* **2020**, *12*, 1354. [CrossRef] [PubMed]
9. Halabi, M.; Mann-Lahav, M.; Beilin, V.; Shter, G.E.; Elishav, O.; Grader, G.S.; Dekel, D.R. Electrospun Anion-Conducting Ionomer Fibers—Effect of Humidity on Final Properties. *Polymers* **2020**, *12*, 1020. [CrossRef] [PubMed]
10. Raja Sulaiman, R.R.; Walvekar, R.; Khalid, M.; Wong, W.Y.; Jagadish, P. Recent Progress in the Development of Aromatic Polymer-Based Proton Exchange Membranes for Fuel Cell Applications. *Polymers* **2020**, *12*, 1061. [CrossRef] [PubMed]

Article

How the Morphology of Nafion-Based Membranes Affects Proton Transport †

Ernestino Lufrano [1], Cataldo Simari [1], Maria Luisa Di Vona [2], Isabella Nicotera [1,*] and Riccardo Narducci [2,*]

[1] Department of Chemistry and Chemical Technologies—CTC, University of Calabria, via Pietro Bucci, 87036 Arcavacata di Rende, Italy; ernestino.lufrano@unical.it (E.L.); cataldo.simari@unical.it (C.S.)
[2] Department of Industrial Engineering and LIME Laboratory, University of Rome Tor Vergata, Via del Politecnico 1, 00133 Rome, Italy; divona@uniroma2.it
* Correspondence: isabella.nicotera@unical.it (I.N.); riccardo.narducci@uniroma2.it (R.N.)
† In memoriam of Prof. Giulio Alberti.

Abstract: This work represents a systematic and in-depth study of how Nafion 1100 membrane preparation procedures affect both the morphology of the polymeric film and the proton transport properties of the electrolyte. The membrane preparation procedure has non-negligible consequences on the performance of the proton-exchange membrane fuel cells (PEMFC) that operate within a wide temperature range (up to 120 °C). A comparison between commercial membranes (Nafion 117 and Nafion 212) and Nafion membranes prepared by three different procedures, namely (a) Nafion-recast, (b) Nafion uncrystallized, and (c) Nafion 117-oriented, was conducted. Electrochemical Impedance Spectroscopy (EIS) and Pulsed-field gradient nuclear magnetic resonance (PFG-NMR) investigations indicated that an anisotropic morphology could be achieved when a Nafion 117 membrane was forced to expand between two fixed and nondeformable surfaces. This anisotropy increased from ~20% in the commercial membrane up to 106% in the pressed membrane, where the ionic clusters were averagely oriented (Nafion 117-oriented) parallel to the surface, leading to a strong directionality in proton transport. Among the membranes obtained by solution-cast, which generally exhibited isotropic proton transport behavior, the Nafion uncrystallized membrane showed the lowest water diffusion coefficients and conductivities, highlighting the correlation between low crystallinity and a more branched and tortuous structure of hydrophilic channels. Finally, the dynamic mechanical analysis (DMA) tests demonstrated the poor elastic modulus for both uncrystallized and oriented membranes, which should be avoided in high-temperature fuel cells.

Keywords: nafion; conductivity; oriented morphology; recast; uncrystallized

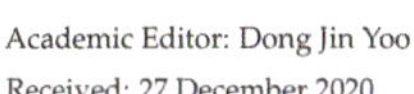

Citation: Lufrano, E.; Simari, C.; Di Vona, M.L.; Nicotera, I.; Narducci, R. How the Morphology of Nafion-Based Membranes Affects Proton Transport†. *Polymers* **2021**, *13*, 359. https://doi.org/10.3390/polym13030359

Academic Editor: Dong Jin Yoo
Received: 27 December 2020
Accepted: 18 January 2021
Published: 22 January 2021

1. Introduction

In recent years, researchers have taken an interest in the development of more sustainable energies, both from an economic and environmental point of view. Among the different types of fuel cells, low-medium temperature, proton-exchange membrane fuel cells (PEMFCs) are promising for the replacement of classic heat engines, especially in motor vehicles [1,2]. Among the most studied and promising materials are perfluorosulfonic acid membranes (PFSA), such as long side chain (LSC) Nafion, which has, until now, been the most widely investigated ionomer, and the more recent short side chain (SSC) Aquivion from Solvay [3,4]. PFSA are characterized by high proton conductivity and chemical inertness; the latter is due to the presence of fluorine. However, sometimes the mechanical and thermal stability are not enough for the present needs in automotive applications [5,6]. In particular, when relative humidity (RH)-temperature conditions overcome certain critical values, (70–130 °C and 95–100% RH [7]), the membranes undergo some irreversible processes that induce a decrease in their through-plane proton conductivity [8]. These phenomena are due to modifications in the bulk-transport properties, and may be observed when a membrane is constrained between the electrodes and forced to swell in a plane direction [9]. A lamellar platelet was first proposed by Fujimura et al. [10] in 1981. In 1982, Starkweather [11] proposed a layered morphology. A lamellar structure was described by Litt et al. [12] in 1997, in which ionic domains were formed out of hydrophilic layers separated by thin lamellar polytetrafluoroethylene (PTFE) crystallites. A lamellar structure was suggested by Haubold et al. [13] in 2001, and recently by Kreuer et al. [14]. Ribbon morphologies were proposed by Gebel [15] in 2000, Rubatat et al. [16], Perrin et al. [17], and more recently, in 2007, by Termonia [18]. In 2013, Alberti and coworkers prepared a large batch of these low-conducting ionomers and hypothesized a layered structure; a change from randomly-oriented to plane-oriented morphologies was proposed to explain the experimental results of low conductivity and low density [19]. The results were interpretated using the INCA method (ionomer n_c analysis) proposed in 2008, and later by Alberti and collaborators [5,20]. This method consists of the elaboration of n_c/T plots, where n_c is the counterpressure index, that describe the counter pressure force of the ionomer matrix that balances the inner osmotic pressure of the proton solution inside the membrane at equilibrium. The use of the n_c/T plot makes it possible to determine the history of the membranes, since the value itself and the trend slope are characteristic of every treatment that the ionomer has undergone: e.g., at 50 °C, the oriented Nafion has a n_c value of approximately two units, much lower than the commercial Nafion 117 membrane, which has a value of about nine units [21,22]. In 2017, some studies, using the Ionomer n_c Analysis (INCA) method, performed on uncrystallized Nafion, with an EW = 1100, and semicrystalline Nafion, with an EW = 1000, evidenced the relation between n_c and the glass transition temperature of the former and the temperature of melting of crystallites in the latter [23]. Using wide-angle (WAXD) and small angle X-ray scattering (SAXS), Moore and Martin [24] found that the as-received and solution-processed films were semicrystalline, with similar degrees of crystallinity, while the recast films were amorphous. The high-solubility and "mud-cracked" character of the recast material suggested that the colloidal morphology observed in the solution remained intact in the recast state, with little chain entanglement or coalescence between particles [1]. As demonstrated by Gebel et al., with WAXS and SAXS in 1987 [25], the polymers threated with a high boiling solvent at a high temperature increased the degree of crystallinity. In 2013, Alberti et al. [26], and in 2018, Narducci et al. [22], reported a thermal treatment assisted by DMSO to increase crystallinity, and proposed a method to avoid or limit low-conductivity phase formation. This method was extended in 2019 by Giancola et al. [4] to Aquivion. A well-defined annealing temperature (T_{ann}), between the glass transition and the melting temperature of the crystalline phase, was chosen for the treatments performed at different times in order to increase the mechanical stability [27].

In order to further confirm and clarify the previous studies carried out using the INCA method, in this study, we analyzed the proton transport behavior by Electrochemical Impedance Spectroscopy (EIS) measurements and Pulse-Field Gradient NMR spectroscopy to obtain direct measurements of the self-diffusion coefficients. Several Nafion-based membranes were investigated and compared: (i) recast Nafion, prepared by a casting procedure from the commercial solution by using DMF as a solvent, (ii) industrial recasting Nafion *NR212*, (iii) oriented Nafion 117, obtained with a special device that mimics the structure of the fuel cell, and (iv) uncrystallized Nafion, obtained by simple evaporation of a commercial solution. All these ionomers were characterized by the same EW = 1100. In particular, our attention was focused on the oriented material, in which through-plane conductivity decay occurred due to the transformation from "random" to "oriented" ribbon-type morphologies, with semicrystalline and amorphous layers being mostly parallel to the surface of the material.

The mechanical properties of such membranes have been investigated by Dynamic Mechanical Analysis (DMA). Particular attention was paid to membranes with low or no crystallinity, i.e., those positioned in the lower part of the n_c/T plot, where the materials were mechanically less stable, and therefore, were similar or inferior to as-received or semicrystalline materials, such as Nafion 117. Once this systematic comparison had been made, we could suggest which materials could be used in fuel cells, and which should be avoided, both for use and formation under operating conditions.

2. Materials and Methods

2.1. Chemicals

Nafion resin solution, 5 wt% solution in a mixture of lower aliphatic alcohols and water (EW, 1100 g eq^{-1}), Nafion 117 membranes (EW, 1100 g eq^{-1}, 180 μm thickness), and other reagents were supplied by Sigma Aldrich (Milan, Italy) and used as received. Nafion NR212 membrane (EW 1100 g eq^{-1}, 51 μm thickness) were supplied by Ion Power Inc. (New Castle, DE, USA) and subjected to thermal and chemical activation before the characterization.

2.2. Preparation of Nafion Uncrystallized

Nafion resin solution was cast on a Petri dish (5 mL), evaporated in air for 24 h at RT, and then placed in an oven for 15 min at 80 °C to eliminate the solvent. The resulting membranes were peeled off the Petri dish and stored in P_2O_5 at RT [23].

2.3. Preparation of Nafion 117 Oriented

Nafion 117 membranes were placed in the apparatus previously described in [19] (Figure 1) with a metallic disc internal diameter of 7 cm. To avoid direct contact with the metal, the plaques were covered with a Teflon foil. The whole apparatus with tightened membranes was placed inside an autoclave at 120 °C, and in liquid water for about 48 h. The pressure in the autoclave simply arose from the vapor pressure of water at 120 °C (i.e., about 2 bar). In this conformation, the membrane swelling perpendicular to the surface was not allowed, and the parallel one was facilitated with the consequent transition to oriented materials. Then, the apparatus was cooled at RT and the samples were maintained in P_2O_5 at RT.

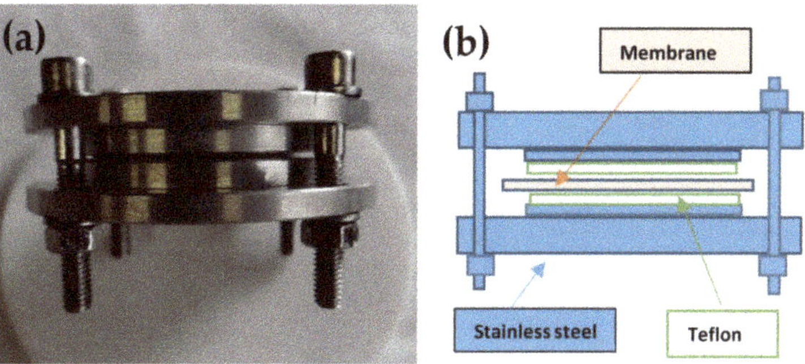

Figure 1. (**a**) Photo and (**b**) schematic representation of the device used for the preparation of Nafion 117 oriented.

2.4. Preparation of Nafion Recast

A Nafion recast membrane was fabricated through a typical solvent casting method. Briefly, 1 g of commercial Nafion perfluorinated resin solution was heated at about 60 °C until the complete evaporation of the solvents (water, alcohol, etc.), and then redissolved in 10 mL of Dimethylformamide (DMF) until a clear solution was obtained. Thereafter, the solution was cast on a Petri dish and placed in the oven at about 60 °C until it was dry. The membrane was finally subjected to thermal and chemical activations according to a standard method [28]. Briefly, for the membrane reinforcement, it was sandwiched and pressed between two Teflon plates and then placed in an oven at 155 °C for 15 min. Thereafter, the membrane was acid-activated by treating it with: (1) 1M HNO_3 solution at 90 °C for 1 h to oxidize the organic impurities, (2) H_2O_2 (3 vol%) at 60 °C for 1 h to remove all the organic impurities, (3) 1M H_2SO_4 at 80 °C for 1 h to remove any metallic impurities, (4) 0.001 M ethylenediaminetetraacetic acid (EDTA) solution at RT for 1 day to remove all the paramagnetic contaminants (such as the copper contained in the Nafion commercial solution), (5) 2 M HCl at 80 °C for 2 h, and (6) EDTA. After each acidic treatment, the membrane was rinsed three times in boiling deionized H_2O for 10 min to remove any trace of the acids. The thickness of the dry membranes was about 50 μm.

2.5. Water Uptake (WU), λ and n_c

The samples were kept in liquid water at 25 °C for 24 h in a hermetically-sealed Teflon vessel. The membranes were carefully wiped off and the mass was determined (m_{wet}). They were then dried over P_2O_5 for 3 days and weighed (m_{dry}). The WU was calculated according to Equation (1):

$$WU = \left(\frac{m_{wet} - m_{dry}}{m_{dry}} \right) \times 100 \quad (1)$$

The hydration number λ, i.e., the number of water molecules per SO_3H group, was calculated by Equation (2):

$$\lambda = \left(\frac{WU}{IEC \times M(H_2O)} \right) \times 10 \quad (2)$$

where the ion exchange capacity (IEC) of Nafion 1100 is 0.909 meq. g^{-1} and M is the molar mass of water. The uncertainty is estimated to be about ± 0.5.

The λ values were converted into n_c values by the Equation (3):

$$n_c = \frac{100}{\lambda - 6} \quad (3)$$

This equation is valid for $\lambda \geq 10$ [5,20].

2.6. ¹H NMR Spectroscopy NMR

The NMR measurements were performed with a Bruker AVANCE 300 Wide Bore NMR spectrometer working at 300 MHz on ^1H, and equipped with a Diff30 Z-diffusion 30 G/cm/A multinuclear probe with exchangeable RF inserts (Bruker, Milan, Italy). The self-diffusion coefficients (D) of water were determined by pulsed field gradient stimulated-echo (PFG-STE) technique [29]. The sequence consists of three 90° rf pulses ($\pi/2$-τ_1-$\pi/2$-τ_m-$\pi/2$) and two gradient pulses applied after the first and the third RF pulse. At time $2\tau_1 + \tau_m$, the echo was found. The FT echo decays were analyzed by means of the relevant Stejskal–Tanner expression Equation (4):

$$I = I_0 \, e^{-\beta D} \quad (4)$$

where I and I_0 represent the intensity/area of a selected resonance peak with and without gradients, respectively, D the self-diffusion coefficient, and β the field gradient parameter. Following the usual notation, the magnetic field pulses had amplitude g, duration d, and time delay Δ. Accordingly, the field gradient parameter can be defined by Equation (5):

$$\beta = \left[(\gamma g \delta)^2 \left(\Delta - \frac{\delta}{3} \right) \right] \quad (5)$$

The used experimental parameters were: δ = 0.8 ms, time delay Δ = 8 ms, and the gradient amplitude varied from 100 to 900 G cm^{-1}. Based on the very low standard deviation of the fitting curve and repeatability of the measurements, the uncertainties of the self-diffusion measurements were approximately 3%. The NMR samples were prepared according to the procedure described in detail elsewhere [30]. The self-diffusion coefficients were measured in the temperature range of 20 °C to 130 °C, measured every 20 °C, leaving the sample to equilibrate at each temperature for approximately 20 min.

2.7. Dynamic Mechanical Analysis DMA

Dynamic Mechanical Analysis (DMA) was conducted by Metravib DMA/25 analyzer equipped with a shear jaw for films clamping (Limonest, France). A dynamic stress of amplitude of 10^{-4} at 1 Hz is applied on a rectangular shaped sample (width = 3 cm; height = 1 cm), in the temperature range of 25–200 °C, with a heating scan rate of 2 °C min^{-1}.

2.8. Electrochemical Impedance Spectroscopy (EIS)

A commercial four-electrode cell (BT-112, Scribner Associates Inc., Southern Pines, NC, USA) was adopted to measure the in-plane proton conductivity of the various Nafion membranes [31]. In this case, the membranes were cut into rectangular shapes of 25 mm × 10 mm. For the through-plane conductivity, the membrane was sandwiched between two disks of conductive carbon papers (d = 10.5 mm) and placed in a homemade two-electrode cell. Impedance spectra were recorded on a PGSTAT30 potentiostat/galvanostat/FRA (Metrohm Autolab B.V., Utrecht, The Netherlands) at OCV, over a frequency range between 1 Hz to 1 MHz, with an oscillating potential of about 10 mV. The resulting impedance data were analyzed by Metrohm Autolab NOVA software. From the Nyquist plot, the electrolyte resistance (R) was extracted as the high-frequency intercept on the real axis, and the ionic conductivity (σ) was calculated according to Equation (6) and reported as an average of three independent measurements:

$$\sigma = \frac{L}{RA} \tag{6}$$

where L is the distance between the electrodes and A is the active area.

The in-plane and through-plane proton conductivities were measured as a function of the temperature, in the range of 20–120 °C, at 90% RH, leaving the sample to equilibrate for at least 30 min before each measurement. A humidification system (Fuel Cells Technologies, Inc Albuquerque, NM, USA) directly connected to the cell was used to finely control temperature and RH.

3. Results and Discussion

3.1. Conductivity Study (Through-Plane vs. In-Plane)

To highlight any sort of anisotropy in the membrane morphology induced by the fabrication procedure, the various Nafion 1100 membranes were investigated by electrochemical impedance spectroscopy (EIS) by using two cell configurations, through-plane (σ_{TP}) and in-plane (σ_{IP}). Figure 2 illustrates the comparison between through-plane and in-plane proton conductivity of the PFSA membranes, in the temperature range of 20 °C to 120 °C at 90% RH. Furthermore, for the sake of comparison among the different Nafion-based membranes, some representative values are also reported in Table 1. In 2008, Holdcroft et al. [32] first compared the in-plane and through-plane conductivity of several Nafion membranes, demonstrating that the conductivity was clearly anisotropic in extruded samples, with the σ_{IP} higher than σ_{TP}, whereas it was isotropic in recast films. Furthermore, a strong correlation between the membrane thickness and the anisotropic degree was also observed: the discrepancy between in-plane and through-plane conductivity was lower for thicker membranes (i.e., 18% for Nafion 117) and increased for thinner membranes, such as Nafion 112, reaching a value of 32%.

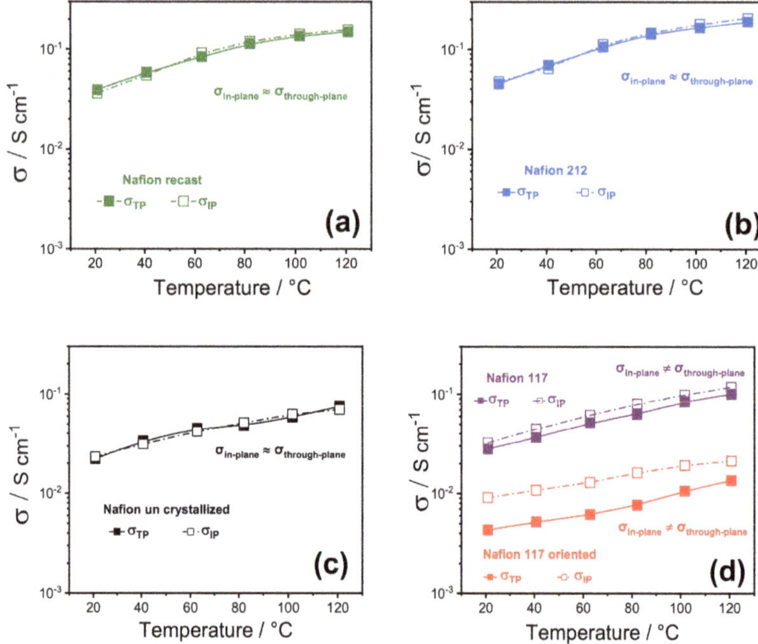

Figure 2. In-plane vs. through-plane proton conductivities of (**a**) Nafion recast, (**b**) Nafion 212; (**c**) Nafion un crystallized and (**d**) Nafion 117 as received and Nafion 117 oriented membranes as a function of temperature (20–120 °C) at 90% RH%. (The error bars are not reported because they are smaller than the size of the symbols).

Table 1. Proton conductivity [mS cm^{-1}], @ RH 90% and for three temperature values, of the PFSA membranes (along with the standard deviation resulting from three independent measurements).

Membranes	40 °C		80 °C		120 °C	
	In-Plane	Through-Plane	In-Plane	Through-Plane	In-Plane	Through-Plane
Nafion 117	40.0 ± 0.8	36.7 ± 1.2	76.5 ± 1.7	63.6 ± 1.3	118.4 ± 1.7	100.6 ± 1.6
Nafion 212	74.1 ± 1.5	76.1 ± 1.6	157.0 ± 2.1	153.1 ± 2.1	220.0 ± 2.1	213.1 ± 2.2
Nafion recast	54.8 ± 1.2	58.7 ± 1.3	119.7 ± 1.9	113.3 ± 1.8	151.0 ± 2.0	149.1 ± 1.9
Nafion uncrystallized	31.2 ± 1.1	30 ± 0.9	52.1 ± 1.1	56.1 ± 1.4	71.3 ± 1.3	76.1 ± 1.6
Nafion 117 oriented	11.07 ± 0.6	5.1 ± 0.3	16.0 ± 0.8	8.08 ± 0.3	21.0 ± 0.8	14.09 ± 0.7

Our findings are consistent with the outcome described by Holdcroft et al., as all the solution-cast membranes (Nafion recast, 212, and uncrystallized) exhibited isotropic conductivity, confirming the absence of any particular orientation in their polymer chains and/or in their ionic clusters. At the same time, a relatively small anisotropy was observed in the case of Nafion 117, with the conductivity parallel to the extruding direction (σ_{IP}) ~ 20% higher than the through-plane. It is worth noting that the in-plane to through-plane anisotropy increased to almost 106% when Nafion 117 oriented. This suggests that the ion-conducting clusters are mostly oriented along the plane of the membrane's surface, inducing a strong directionality to the proton transport. This is expected to greatly affect the performance of the electrolyte in the PEMFC, where the σ_{TP} is crucial. In particular, while the through-plane conductivity of Nafion 212, recast, 117, and partially uncrystallized membranes still match the requirements for PEMFC application, the low σ_{TP} of Nafion 117 oriented makes it unsuitable for practical application; for this reason, its formation in the fuel cell itself should also be avoided. Such a huge decrease in the through-plane conductivity of Nafion 117 oriented is also accompanied by an appreciable decrease of the ionomer density. It is found to be 1.7–1.8 g cm^{-3}, a little higher than that found in the previous work, i.e., 1.4–1.5 g cm^{-3}, probably due to some small differences in the procedure, and because the previous device had a smaller diameter than the one used here. However, such a density reduction is sufficient to cause the drop in conductivity, that we attribute to the transition of ribbon phases already present in commercial Nafion, from a random orientation to parallel direction of the surface, probably causing less/worsened interconnection between the ion channels, giving rise to lower through-plane conductivity.

3.2. *Hydration Number (λ) and Counterpressure Index (n_c)*

From the water uptake measured for each membrane and reported in Table 2, it was possible to calculate λ and n_c. Based on these values, we can determine, without building the whole n_c/T plots at different temperatures, the position in the plot at 25 °C of the single sample, suggesting the type of Nafion we are investigating and what kind of treatment it has previously undergone.

Table 2. Thickness, WU%, lambda, and nc for Nafion 117, NR212, "recast", "uncrystallized" and "oriented" at 25 °C in liquid water for 24 h.

Membranes	Thickness [µm]	WU [wt%]	λ	n_c
Nafion 117	180 ± 6	24	14.6	11.6
Nafion 212	51 ± 2	22	13.4	13.5
Nafion recast	50 ± 1	24	14.6	11.6
Nafion uncrystallized	59 ± 2	29	17.7	8.6
Nafion 117 oriented	160 ± 5	25	15.3	10.8

The values of n_c fall with good approximation in Figure 3 of the reference 17, suggesting the good reproducibility of the system [19].

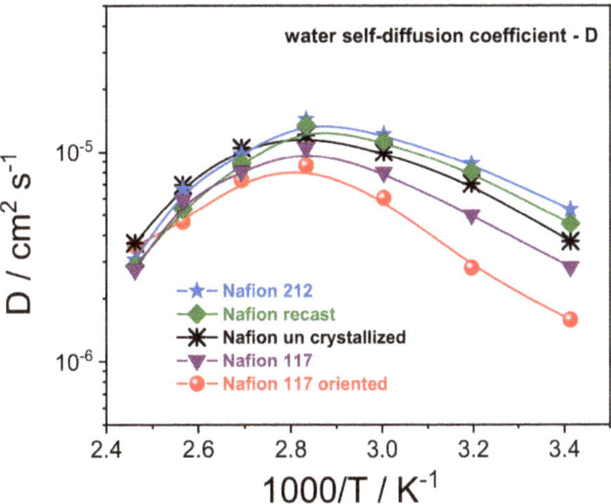

Figure 3. Self-diffusion coefficients as a function of the temperature (from 20 °C to 130 °C) of the water confined in the various PFSA membranes.

From Table 2, Nafion 212 has the lowest water uptake and the highest value of n_c. One need to consider that, according to Alberti et al. [5], one n_c unit is equivalent to an increase of Young's modulus by about 6.5 MPa, with the higher the value having the greater mechanical resistance.

The highest value of WU was reached for the uncrystallized membrane. The thickness of this sample is comparable to commercial Nafion 212, i.e., about 50 μm. Both membranes have been obtained by recasting method; however, the preparation of the uncrystallized does not use a high-boiling solvent, in contrast to Nafion 212. In fact, as reported in the literature (e.g., Alberti [26], Gebel [25], Moore [24]), the use of high-boiling solvents such as DMF and DMSO increases the crystallinity on both preformed and casting membranes. Therefore, the uncrystallized Nafion membrane demonstrated a weaker structure. Finally, the Nafion 117 oriented presented a lower density, had a higher equivalent volume, and therefore, had a greater willingness to host water [19].

H PFG NMR Investigation

^{1}H-PFG NMR experiments allowed us to investigate the molecular dynamics of the water confined inside the prepared electrolytes, via the direct measurements of the water self-diffusion coefficients (D) [33,34]. Figure 3 shows the Arrhenius plot of the water self-diffusion coefficients measured on completely swelled membranes in the temperature range between 20 and 130 °C. For all the membranes, the diffusivity increases with temperature up to 80 °C due to thermal energy, then drops precipitously as a consequence of the evaporation of the bulk-like water fraction. In this regard, once above 80 °C, the D values of the various Nafion membranes become quite comparable and merge towards the same value. This is indicative that the "bound-water" fraction that remains in the ion clusters as hydration to sulfonic groups is almost invariable, regardless of the casting procedure/preparation of the polymeric film. Instead, the analysis of D in the temperature range of 20–80 °C allows us to clarify the relationship between proton transport properties and a membrane's nanomorphology:

Despite the highest water uptake (29 wt%), uncrystallized Nafion displays slightly lower D than both Nafion 212 and Nafion recast (*w.u.* 22 wt% and 24 wt%, respectively). Likely, the decreasing of the crystallinity degree induces a more branched channel structuring, increasing the diffusion path's tortuosity for water, thus slowing the overall diffusivity.

A Nafion membrane structure more regular with a lamellar morphology of the polymer chains reduces the mobility of water molecules, being limited to only two directions. Indeed, Nafion 117 and, even more, the Nafion 117 oriented, show a significant decrease of water diffusion.

3.3. Dynamic Mechanical Analysis

Figure 4 shows the temperature evolution of storage modulus (E') and dumping factor (tan δ) of all the membranes. It can be observed that the storage moduli of Nafion 212 and Nafion recast (both thermally activated before the characterization) are almost comparable, and are the highest among the investigated membranes: E' for these membranes is ca. 200 MPa, which is almost 3.5-fold higher than the other samples in which the storage modulus is ca. 60 MPa.

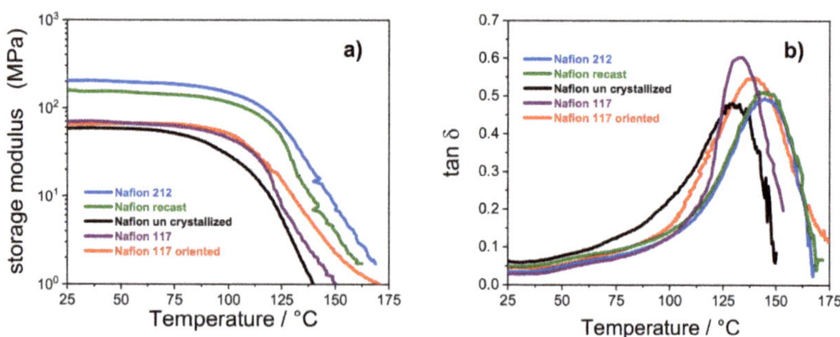

Figure 4. Storage modulus (**a**) and tan δ (**b**) versus temperature of PFSA membranes.

This outcome confirms that the thermal activation procedure is crucial to achieving tough and mechanically stable polymeric film. As noted in Table 2, the Nafion 212 and recast PEMs show a high value of n_c, which, according to Alberti et al. [5], is related to a greater mechanical stability. It is worth noting that the storage modulus of uncrystallized Nafion starts decreasing at 70 °C, whereas the other membranes show a high E′ until 100 °C. As described in the experimental methods, the Nafion uncrystallized was prepared in absence of high-boiling solvents. This reduces the overall crystallinity of the membrane and clearly has a detrimental effect on its thermal resistance. The dumping factor plots showed in Figure 4b further elucidates the relationship between thermal resistance and membranes architecture. The tan δ profile of each Nafion membrane is characterized by a single peak in the high temperature region that is typically related to the α transition (T_g) of the ionic clusters [35,36]. In the case of solution-cast membranes, the T_g of uncrystallized Nafion is shifted to lower temperatures (i.e., 128 °C) in comparison with Nafion 212 and Nafion recast, which show the α transition at ~145 °C. The outcome, which is clearly amenable to the lower crystallinity of the uncrystallized Nafion, de facto limits the practical application of this PEM in high-temperature fuel cell devices, which has a temperature target fixed by DOE of 120 °C. Turning our attention to the extruded membranes, the higher T_g of Nafion 117 oriented compared to the parental Nafion 117 (i.e., 139 °C vs. 132 °C, respectively) is compatible with a reduced flexibility of the polymer chains after strict orientation of its ionic clusters, which need more energy to move, leading to a higher relaxation temperature.

4. Conclusions

This study extends our knowledge of the morphology of Nafion membranes from previous INCA method results. In particular, it has been argued that the decay of the through-plane conductivity in Nafion-oriented membranes is the result of the formation of "oriented-ribbon"-type morphologies and poor connections between ion clusters, which macroscopically is accompanied by a consistent decrease in density. In this study, both in-plane and through-plane conductivities were investigated using a wide range of temperatures (20–120 °C). There was evidence of a strong anisotropy between σ_{TP} and σ_{IP} in the case of extruded membranes, which increases from 20% for the Nafion 117 commercial membrane to almost 106% in the Nafion 117-oriented membrane. In this last sample, the lamellar-like morphology significantly reduced the water diffusivity because it was restricted in only two directions.

Regarding the uncrystallized Nafion, in previous studies, we observed a strong lowering of n_c/T plots due to either an important decreasing in crystallinity, or more likely due to a decomposition of the hydrogen bonds between adjacent ribbons. The high water uptake ability of this membrane, however, is not accompanied by a corresponding high conductivity. On the contrary, both the lower proton conductivity and water diffusion, compared to Nafion 212 and Nafion recast, suggest that the decreased degree of crystallinity induces more branched channel structuring, increasing tortuosity and slowing down the overall transport properties. In addition, the β-transition shifts at 20 °C lower than the other solution-cast membranes, which, along with having a lower storage modulus, definitely limits the practical application of uncrystallized Nafion. On the other hand, materials with a high degree of crystallinity induced by the use of high-boiling solvents, such as DMF and DMSO, have optimal structural characteristics for use in medium/high-temperature fuel cells.

Author Contributions: Conceptualization, R.N. and I.N.; methodology, R.N., I.N. and M.L.D.V.; software, C.S. and E.L.; Writing-Original Draft Preparation, E.L. and R.N.; Writing-Review & Editing, C.S. and I.N.; validation, R.N. and E.L. All authors have read and agreed to the published version of the manuscript.

Funding: This research received no external funding.

Acknowledgments: The authors are indebted to the late G. Alberti (May 1930–October 2017) for having initiated this study.

Conflicts of Interest: The authors declare no conflict of interest.

References

1. Mauritz, K.A.; Moore, R.B. State of understanding of nafion. *Chem. Rev.* **2004**, *104*, 4535–4585. [CrossRef] [PubMed]
2. Alberti, G.; Casciola, M.; Donnadio, A.; Narducci, R.; Pica, M.; Sganappa, M. Preparation and properties of nafion membranes containing nanoparticles of zirconium phosphate. *Desalination* **2006**, *199*, 280–282. [CrossRef]
3. Kusoglu, A.; Weber, A.Z. New Insights into Perfluorinated Sulfonic-Acid Ionomers. *Chem. Rev.* **2017**, *117*, 987–1104. [CrossRef] [PubMed]
4. Giancola, S.; Arciniegas, R.A.B.; Fahs, A.; Chailan, J.F.; Di Vona, M.L.; Knauth, P.; Narducci, R. Study of annealed aquivion® ionomers with the INCA method. *Membranes* **2019**, *9*, 1–13. [CrossRef] [PubMed]
5. Alberti, G.; Narducci, R.; Sganappa, M. Effects of hydrothermal/thermal treatments on the water-uptake of Nafion membranes and relations with changes of conformation, counter-elastic force and tensile modulus of the matrix. *J. Power Sources* **2008**, *178*, 575–583. [CrossRef]
6. Kreuer, K.D. The role of internal pressure for the hydration and transport properties of ionomers and polyelectrolytes. *Solid State Ion.* **2013**, *252*, 93–101. [CrossRef]
7. Alberti, G.; Casciola, M. Composite Membranes for Medium -Temperature Pem Fuel Cells. *Annu. Rev. Mater. Sci.* **2003**, *33*, 129–154. [CrossRef]
8. Casciola, M.; Alberti, G.; Sganappa, M.; Narducci, R. Factors affecting the stability of Nafion conductivity at high temperature and relative humidity. *Desalination* **2006**, *200*, 639–641. [CrossRef]
9. Casciola, M.; Alberti, G.; Sganappa, M.; Narducci, R. On the decay of Nafion proton conductivity at high temperature and relative humidity. *J. Power Sources* **2006**, *162*, 141–145. [CrossRef]
10. Fujimura, M.; Hashimoto, T.; Hawai, H. Small-Angle X-ray Scattering Study of Perfluorinated Ionomer Membranes. 1. Origin of Two Scattering Maxima. *Macromolecules* **1981**, *14*, 1309–1315. [CrossRef]
11. Starkweather, H.W. Crystallinity in Perfluorosulfonic Acid Ionomers and Related Polymers. *Macromolecules* **1982**, *15*, 320–323. [CrossRef]
12. Litt, M.H. Reevaluation of Nafion morphology. *Polym. Prepr.* **1997**, *213*, 80–81.
13. Haubold, H.G.; Vad, T.; Jungbluth, H.; Hiller, P. Nano structure of NAFION: A SAXS study. *Electrochim. Acta* **2001**, *46*, 1559–1563. [CrossRef]
14. Kreuer, K.D.; Portale, G. A critical revision of the nano-morphology of proton conducting ionomers and polyelectrolytes for fuel cell applications. *Adv. Funct. Mater.* **2013**, *23*, 5390–5397. [CrossRef]
15. Gebel, G. Structural evolution of water swollen perfluorosulfonated ionomers from dry membrane to solution. *Polymer* **2000**, *41*, 5829–5838. [CrossRef]
16. Rubatat, L.; Gebel, G.; Diat, O. Fibrillar structure of Nafion: Matching fourier and real space studies of corresponding films and solutions. *Macromolecules* **2004**, *37*, 7772–7783. [CrossRef]
17. Perrin, J.C.; Lyonnard, S.; Guillermo, A.; Levitz, P. Water dynamics in ionomer membranes by field-cycling NMR relaxometry. *J. Phys. Chem. B* **2006**, *110*, 5439–5444. [CrossRef]
18. Termonia, Y. Nanoscale modeling of the structure of perfluorosulfonated ionomer membranes at varying degrees of swelling. *Polymer* **2007**, *48*, 1435–1440. [CrossRef]
19. Alberti, G.; Narducci, R.; Di Vona, M.L.; Giancola, S. More on Nafion conductivity decay at temperatures higher than 80 °C: Preparation and first characterization of in-plane oriented layered morphologies. *Ind. Eng. Chem. Res.* **2013**, *52*, 10418–10424. [CrossRef]
20. Alberti, G.; Narducci, R. Evolution of permanent deformations (ormemory) in nafion 117membranes with changes in temperature, relative humidity and time, and its importancein the development of medium temperature PEMFCs. *Fuel Cells* **2009**, *9*, 410–420. [CrossRef]
21. Alberti, G.; Di Vona, M.L.; Narducci, R. New results on the visco-elastic behaviour of ionomer membranes and relations between T-RH plots and proton conductivity decay of Nafion ® 117 in the range 50–140 °C. *Int. J. Hydrogen Energy* **2012**, *37*, 6302–6307. [CrossRef]
22. Narducci, R.; Knauth, P.; Chailan, J.F.; Di Vona, M.L. How to improve Nafion with tailor made annealing. *RSC Adv.* **2018**, *8*, 27268–27274. [CrossRef]
23. Alberti, G.; Narducci, R.; Di Vona, M.L.; Giancola, S. Preparation and Nc/T plots of un-crystallized Nafion 1100 and semi-crystalline Nafion 1000. *Int. J. Hydrogen Energy* **2017**, *42*, 15908–15912. [CrossRef]
24. Moore, R.B.; Martin, C.R. Chemical and Morphological Properties of Solution-Cast Perfluorosulfonate Ionomers. *Macromolecules* **1988**, *21*, 1334–1339. [CrossRef]
25. Gebel, G.; Aldebert, P.; Pineri, M. Structure and Related Properties of Solution-Cast Perfluorosulfonated Ionomer Films. *Macromolecules* **1987**, *20*, 1425–1428. [CrossRef]
26. Alberti, G.; Narducci, R.; Di Vona, M.L.; Giancola, S. Annealing of nafion 1100 in the presence of an annealing agent: A powerful method for increasing ionomer working temperature in PEMFCs. *Fuel Cells* **2013**, *13*, 42–47. [CrossRef]
27. Young, R.J.; Lovell, P.A. *Introduction to Polymers*, 3rd ed.; CRC Press, Taylor and Francis Group: Boca Raton, FL, USA, 2011, ISBN 9781439894156.

28. Enotiadis, A.; Boutsika, L.G.; Spyrou, K.; Simari, C.; Nicotera, I. A facile approach to fabricating organosilica layered material with sulfonic groups as an efficient filler for polymer electrolyte nanocomposites. *New J. Chem.* **2017**, *41*, 9489–9496. [CrossRef]
29. Tanner, J.E. Use of the stimulated echo in NMR diffusion studies. *J. Chem. Phys.* **1970**, *52*, 2523–2526. [CrossRef]
30. Nicotera, I.; Simari, C.; Boutsika, L.G.; Coppola, L.; Spyrou, K.; Enotiadis, A. NMR investigation on nanocomposite membranes based on organosilica layered materials bearing different functional groups for PEMFCs. *Int. J. Hydrogen Energy* **2017**, *42*, 27940–27949. [CrossRef]
31. Simari, C.; Stallworth, P.; Peng, J.; Coppola, L.; Greenbaum, S.; Nicotera, I. Graphene oxide and sulfonated-derivative: Proton transport properties and electrochemical behavior of Nafion-based nanocomposites. *Electrochim. Acta* **2019**, *297*, 240–249. [CrossRef]
32. Soboleva, T.; Xie, Z.; Shi, Z.; Tsang, E.; Navessin, T.; Holdcroft, S. Investigation of the through-plane impedance technique for evaluation of anisotropy of proton conducting polymer membranes. *J. Electroanal. Chem.* **2008**, *622*, 145–152. [CrossRef]
33. Simari, C.; Lufrano, E.; Coppola, L.; Nicotera, I. Composite gel polymer electrolytes based on organo-modified nanoclays: Investigation on lithium-ion transport and mechanical properties. *Membranes* **2018**, *8*, 69. [CrossRef] [PubMed]
34. Simari, C.; Baglio, V.; Lo Vecchio, C.; Aricò, A.S.; Agostino, R.G.; Coppola, L.; Oliviero Rossi, C.; Nicotera, I. Reduced methanol crossover and enhanced proton transport in nanocomposite membranes based on clay−CNTs hybrid materials for direct methanol fuel cells. *Ionics* **2017**, *23*. [CrossRef]
35. Mauritz, K.A.; Stefanithis, I.D. Microstructural Evolution of A Silicon Oxide Phase in a Perfluorosulfonic Acid Ionomer by an in Situ Sol-Gel Reaction. 2. Dielectric Relaxation Studies. *Macromolecules* **1990**, *23*, 1380–1388. [CrossRef]
36. Kyu, T.; Eisenberg, A. *In Perfluorinated Ionomer Membranes*; American Chemical Society: Washington, DC, USA, 1982; Volume 6, pp. 93–100, ISBN 0841206988.

Article

Stability of Proton Exchange Membranes in Phosphate Buffer for Enzymatic Fuel Cell Application: Hydration, Conductivity and Mechanical Properties

Luca Pasquini [1,*], Botagoz Zhakisheva [1], Emanuela Sgreccia [2], Riccardo Narducci [2], Maria Luisa Di Vona [2] and Philippe Knauth [1]

[1] CNRS, MADIREL (UMR 7246) and International Laboratory: Ionomer Materials for Energy, Aix Marseille Univ, Campus St. Jérôme, 13013 Marseille, France; botagoz.zhakisheva@etu.univ-amu.fr (B.Z.); philippe.knauth@univ-amu.fr (P.K.)
[2] Department Industrial Engineering and International Laboratory: Ionomer Materials for Energy, University of Rome Tor Vergata, 00133 Roma, Italy; emanuela.sgreccia@uniroma2.it (E.S.); riccardo.narducci@uniroma2.it (R.N.); divona@uniroma2.it (M.L.D.V.)
* Correspondence: luca.pasquini@univ-amu.fr

Abstract: Proton-conducting ionomers are widespread materials for application in electrochemical energy storage devices. However, their properties depend strongly on operating conditions. In bio-fuel cells with a separator membrane, the swelling behavior as well as the conductivity need to be optimized with regard to the use of buffer solutions for the stability of the enzyme catalyst. This work presents a study of the hydrolytic stability, conductivity and mechanical behavior of different proton exchange membranes based on sulfonated poly(ether ether ketone) (SPEEK) and sulfonated poly(phenyl sulfone) (SPPSU) ionomers in phosphate buffer solution. The results show that the membrane stability can be adapted by changing the casting solvent (DMSO, water or ethanol) and procedures, including a crosslinking heat treatment, or by blending the two ionomers. A comparison with Nafion™ shows the different behavior of this ionomer versus SPEEK membranes.

Keywords: ionomer; blend; casting; SPEEK; SPPSU; crosslinking

Citation: Pasquini, L.; Zhakisheva, B.; Sgreccia, E.; Narducci, R.; Di Vona, M.L.; Knauth, P. Stability of Proton Exchange Membranes in Phosphate Buffer for Enzymatic Fuel Cell Application: Hydration, Conductivity and Mechanical Properties. *Polymers* **2021**, *13*, 475. https://doi.org/10.3390/polym13030475

Academic Editor: Dong Jin Yoo

Received: 3 January 2021
Accepted: 27 January 2021
Published: 2 February 2021

1. Introduction

Synthetic polymer membranes are economical separator materials in many advanced devices, including ultrafiltration systems, electrolyzers, redox-flow batteries and hydrogen separation [1–5]. Membranes composed of ionomers [6] (i.e., polymers with grafted ionic groups, including commercial Nafion™, 3M™, Aquivion™) are important for the development of highly efficient and reliable electrochemical energy conversion devices such as fuel cells. The main challenge is to increase the ionomer conductivity and maintain an appropriate hydrolytic and mechanical stability [7–9]; very often, the good conductivity of highly functionalized polymers is linked to a poor hydrolytic stability [10,11], an increase of the reactant permeability and an overall decay of the device performances, ultimately causing the failure of the systems. The microstructure and behavior of ionomer membranes in operating conditions thus need to be better understood to increase the lifetime and the efficiency of the devices.

In the last years, biological fuel cells (BioFCs) were developed as an alternative to classical fuel cells [12]. The main advancement and, at the same time, the main challenge for this type of cell is the substitution of the inorganic catalyst. Generally, expensive and rare (e.g., platinum) [13,14] BioFCs can be divided in two sub-categories [15], microbial fuel cells (MFCs), using complex organisms as catalyst [16,17], and enzymatic fuel cells (EFCs), using enzymes as catalyst [18–20]. In both cases, the coupling of the catalyst with the materials of the whole device has to be deeply understood to guarantee the catalyst's stability over time and the overall performance to the device [14]. EFCs including a separator membrane [20,21] were introduced to solve the main problems of membrane-less EFC, i.e., the mixing of the reactants (O_2 and H_2) and the sensitivity of some enzymes to O_2 [22], which result in a short lifetime of the device and poor performances [23,24]. In the specific example of the EFCs, the enzyme catalyst is often immersed in a buffer solution to guarantee the optimal activity [24]. All materials utilized are thus in contact with the buffer, including the ionomer membrane that has to be hydrolytically stable and maintain a sufficient ionic conductivity in this medium.

We already demonstrated [25] that the hydrolytic stability and conductivity of proton and anion conducting membranes strongly depends on the type, concentration and pH of the buffer: ionic crosslinks by ions contained in the solution can significantly affect the conductivity. Some previous works [26–30] demonstrate that commercially available Nafion™ membranes can undergo a dramatic proton conductivity reduction in BioFCs related to cation exchange. After a few hours of EFC operation, the membrane completely exchanged cations and required an intensive wash with sulfuric acid or a replacement, increasing the overall maintenance cost of the device. It is thus of crucial importance to study the behavior of ionomers in typical EFC conditions and in particular how the hydrolytic stability, as well as the conductivity, can be optimized by different methods (such as blend formation between polymers, change of the casting procedure and solvent, crosslinking reactions) to boost the overall performances of the device.

In this work, we investigate the casting of two main sulfonated aromatic polymers (SAP), sulfonated poly(ether ether ketone) (SPEEK) and sulfonated poly(phenyl sulfone) (SPPSU), as shown in Scheme 1a, and their blends in different solvents to obtain an optimized membrane for the EFC phosphate buffer solution. In particular, the casting of highly functionalized SPEEK from dimethylsulfoxide (DMSO), DMSO-water, ethanol and ethanol-water, and the blending of SPEEK and SPPSU, are investigated to optimize the conductivity of the membranes while maintaining a good hydrolytic and mechanical stability. The properties of membranes are reported after immersion in a 0.05 M phosphate buffer solution, including conductivity, hydrolytic stability (gravimetric and volumetric solution uptake), dry density and Young's modulus. These properties are finally compared to commercial Nafion™ 212 that is used as a benchmark.

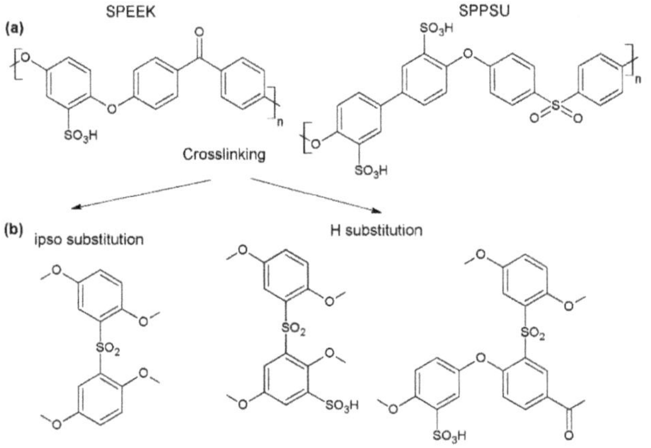

Scheme 1. (**a**) Structures of sulfonated poly(ether ether ketone) (SPEEK) and sulfonated poly(phenyl sulfone) (SPPSU). (**b**) Various possible pathways for the reticulation of SPEEK.

2. Materials and Methods

The Nafion™ 212 membrane was purchased from Fuel Cell Store (College Station, TX, USA) in the form of a film of 30 × 30 cm^2 dimension.

Sulfonated poly(ether ether ketone) (SPEEK) and sulfonated poly(phenyl sulfone) (SPPSU) were prepared by reaction of poly(ether ether ketone) (Victrex, Thornton-Cleveleys,

UK, MW = 38,300 g/mol) or poly(phenyl sulfone) (Solvay, Brussels, Belgium, MW = 46,173 g/mol) with concentrated sulfuric acid (H_2SO_4; Aldrich, St. Louis, MO, USA, 95–97%) under nitrogen atmosphere [31].

The degree of sulfonation (DS) and the ion exchange capacity (IEC) of each polymer was determined by NMR spectroscopy and acid-base titration [32,33]. The obtained values were, for SPEEK: DS = 92% and IEC = 2.50 meq/g, and for SPPSU: DS = 152% and IEC = 2.92 meq/g.

Membranes were cast from DMSO, DMSO-water, ethanol or ethanol-water in a flat Petri dish (Figure 1). Generally, 0.5 g of polymer (or a mixture of two polymers) was dissolved in 10 g of solvent (or a mixture of two solvents). In the case of DMSO or DMSO-water, after evaporation to around one third of the original volume, the solution was poured in a Petri dish and evaporated in an oven at 80 °C for 18 h.

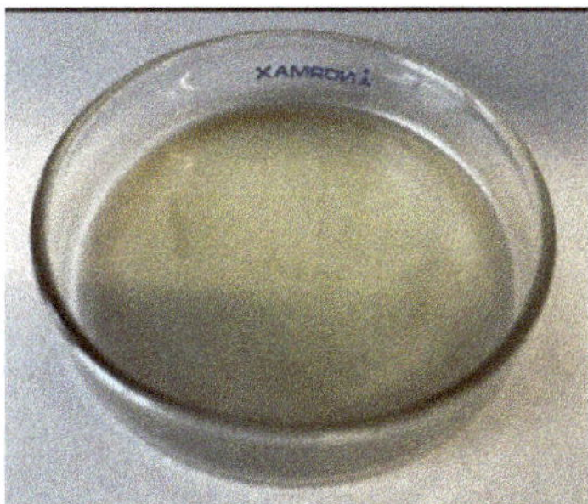

Figure 1. A typical membrane cast in a flat Petri dish.

For ethanol and ethanol-water, the solution was directly poured into a Petri dish and evaporated in an oven at 80 °C for 15 min or 18 h, respectively.

A crosslinking treatment at 180 °C for 3 h, as described in References [11,34,35], was applied to as-cast membranes that showed poor hydrolytic stability (in particular, SPPSU-based membranes).

2.1. Hydrolytic Stability

The hydrolytic stability was investigated in a 0.05 M phosphate ($H_2PO_4^-/HPO_4^{2-}$) buffer at pH = 6.5 ± 0.2. The pH was determined with a calibrated pH-meter (Mettler Toledo). This buffer was chosen because it is among the best buffer solutions that stabilize most of the enzymes in EFCs, in particular bilirubin oxidase [12,36].

Mass uptake (*MU*) was measured twice at 25 °C and calculated according to the equation:

$$MU(\%) = \frac{m_{wet} - m_{dry}}{m_{dry}} a \times 100 \quad (1)$$

The mass of wet samples (m_{wet}) was determined after immersion in the buffer solution during 24 h at 25 °C in a thermoregulated oven without any additional washing in water. Before the measurement, the membrane was wiped carefully with absorbing paper to remove the excess of buffer solution on the surface. The mass of the dry samples (m_{dry}) was measured in a closed vessel after drying over P_2O_5 for 24 h.

The dry density of the ionomer was measured using the mass and dimensions of the membranes after drying over P_2O_5 for 24 h.

2.2. Ionic Conductivity

The through-plane ionic conductivity was measured by impedance spectrometry between 1 Hz and 6 MHz using an impedance spectrometer, Biologic VSP300 (Biologic, Seyssinet-Pariset, France). The amplitude of the oscillating voltage was 20 mV. After immersion in the buffer during 24 h at 25 °C in a thermoregulated oven and after removing the buffer excess on the surface, the samples were measured at 25 °C in humidified conditions inside a Swagelok cell with two stainless-steel electrodes. The sample resistance, R, was obtained from typical impedance spectra (Figure 2) using the intercept with the real axis. The ionic conductivity, σ, was calculated using the equation:

$$\sigma = \frac{th_{wet}}{R\ a \cdot\ A_{wet}} \quad (2)$$

where th_{wet} and A_{wet} are respectively the thickness of the membrane in the wet state after the measurement (measured with a micrometer, Mitutoyo 293-230) and the electrode area.

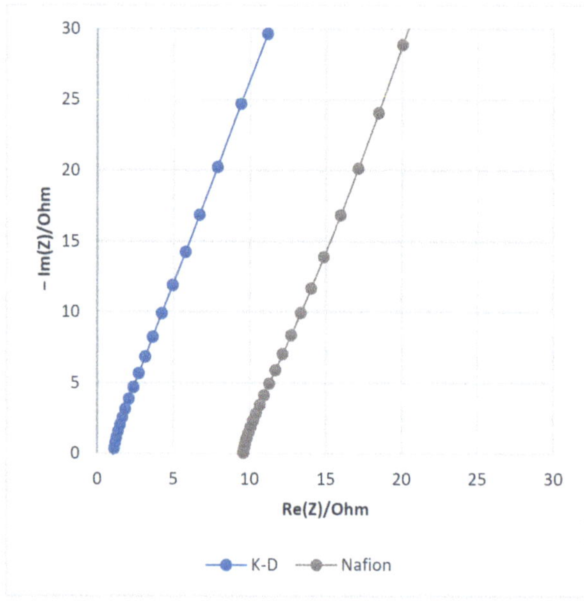

Figure 2. Typical complex impedance diagrams of a SPEEK membrane cast in DMSO (K-D) and a Nafion^{TM} membrane after immersion in buffer.

2.3. Tensile Stress-Strain Tests

The mechanical measurements were performed with an Adamel Lhomargy TESTOMETRIC M250-2.5CT testing machine (Testometric, Rochdale, UK). The specimens (two for each test) were cut in rectangular shape (25 mm length and 5 mm width) by paying particular attention to having smooth and perfectly sharp edges to avoid points of stress concentration that eventually lead to premature breaking. The samples were placed between two clamps and the test was performed at a constant elongation rate of 5 mm/min at ambient temperature (25 ± 1 °C) and relative humidity (RH 40% ± 10%), or after immersion at 25 °C in the phosphate buffer solution.

3. Results

3.1. Cast Membranes

The casting conditions of the realized membranes are reported in Table 1.

Table 1. Casting conditions of membranes. K stands for SPEEK, U for SPPSU, D for DMSO, W for water, and E for ethanol. XL (crosslink) indicates membranes thermally treated 3 h at 180 °C. sh specifies a short (15 min) heat treatment during casting. The casting temperature in all cases is 80 °C.

Name	Polymer (wt %)		Casting Solvent (wt %)			Casting Time
	SPPSU	SPEEK	DMSO	Water	Ethanol	
K-D	-	100	100	-	-	18 h
K-E-sh	-	100	-	-	100	15 min
K-E	-	100	-	-	100	18 h
K-EW-sh	-	100	-	50	50	15 min
K-EW	-	100	-	50	50	18 h
U-D	100	-	100	-	-	18 h
KU50-D	50	50	100	-	-	18 h
KU30-D	30	70	100	-	-	18 h
U-DW	100	-	10	90	-	18 h
KU50-DW (XL)	50	50	10	90	-	18 h
KU30-DW (XL)	30	70	10	90	-	18 h
U-E-sh	100	-	-	-	100	15 min
U-E	100	-	-	-	100	18 h

The various polymer membranes were realized to study how the casting procedure affected the membrane behavior in the buffer solution. In particular, the utilization of solvents different from DMSO was explored to cast membranes free of harmful solvent for enzymes. The casting temperature of 80 °C was chosen following our previous experience with membranes as energy devices [33,37,38]: it is compatible both with DMSO and ethanol and in the case of ethanol, avoids a too fast evaporation resulting in an inhomogeneous membrane.

SPEEK/SPPSU blends were explored to simultaneously enhance the conductivity and hydrolytic stability of the membrane: SPEEK contributed to the hydrolytic stability and SPPSU contributed to increase the ionic conductivity. Following the results of the characterization of SPEEK/SPPSU blends, we moved from the initial 50/50 wt % to a 70/30 wt % composition, because of the poor stability in the buffer of the first blend (cf. Discussion, Section 4).

The crosslinked membranes (indicated with XL) were realized with the aim of hydrolytic stabilization even with a small loss of ionic conductivity because of the decrease of the ion exchange capacity (IEC) due to the reticulation reaction [35,39]. As described in Scheme 1b, the crosslinking treatment consumes some sulfonic acid groups for the creation of sulfone bridges between chains: following XL, the polymer absorbs less water, because of the loss of some hydrophilic sulfonic acid groups resulting in an increase of the hydrolytic stability. A better compromise between stability and good ionic conductivity is thus possible to reach by reticulation.

The hydrolytic stability and conductivity in phosphate buffer solution, as well as the dry density and Young's modulus of membranes, are reported in Table 2.

The dry density of membranes depends strongly on the casting solvent. The dry density of SPEEK and SPPSU has the highest value of around 1.4 g/cm^3 when cast from DMSO, in good agreement with literature values [9,40]. This result indicates a high packing density of the macromolecules, also increasing the stiffness of the membranes by Van der Waals interactions (see below). SPEEK membranes cast from ethanol present a lower density, around 1 g/cm^3, indicating that in this solvent, the macromolecular chains are much less densely packed with no nanophase separation (cf. Discussion, Section 4). Intermediate values, more similar to DMSO, are observed in ethanol/water mixtures. In the case of SPPSU and SPPSU/SPEEK blends, the dry density is around 1.2 g/cm^3 and increases as expected after a cross-linking treatment.

Table 2. Mass (MU) and volume (VU) uptake, conductivity (σ), dry density (d) and Young's modulus (E) in humid air of various ionomer membranes. K stands for SPEEK, U for SPPSU, D for DMSO, W for water, and E for ethanol. XL (crosslink) indicates membranes thermally treated for 3 h at 180 °C. sh specifies a short (15 min) heat treatment during casting.

Samples	MU (%)	VU (%)	σ (mS/cm)	d (g/cm^3)	E (MPa)
K-D	74.3	115.9	15.8	1.44	1400 ± 150
K-E-sh	60.7	98.8	9.0	1.09	900 ± 160
K-E	67.0	110.9	12.7	0.99	580 ± 10
K-EW-sh	279.1	419.7	22.2	1.31	1370 ± 10
K-EW	256.1	382.8	22.9	1.39	1090 ± 5
U-D	Diss *	Diss *	Diss *	1.32	700 ± 50
KU50-D	partial diss	not measured	25.9	1.19	370 ± 80
KU30-D	91	176.4	25.8	1.06	640 ± 10
U-DW	Diss *	Diss *	Diss *	1.15	590 ± 280
KU50-DW	Diss *	Diss *	Diss *	1.13	1220 ± 10
KU30-DW	174	214.4	28.8	1.15	830 ± 120
KU50-DW XL	79.7	192.0	10.4	1.61	not measured
KU30-DW XL	74.4	80.8	10.3	1.27	not measured
U-E-sh	Diss *	Diss *	Diss *	1.23	830 ± 120
U-E	Diss *	Diss *	Diss *	1.20	500 ± 40

* Dissolution.

3.2. MU, Dry Density, Conductivity, Mechanical Properties

Figure 2 shows typical impedance spectra of SPEEK (K-D) and Nafion^{TM} membranes.

3.3. Comparison of SPEEK vs. Nafion^{TM}

The MU, conductivity and mechanical properties were compared for a typical DMSO-cast SPEEK membrane (K-D) and a Nafion^{TM} 212 membrane before and after immersion in buffer solution.

Figure 3 presents typical stress-strain curves for SPEEK and Nafion^{TM}. The mechanical properties are reported in Table 3. The curves and properties in ambient humidity are consistent with previously reported data [40,41]. After immersion in buffer solutions, the curves change dramatically. Whereas the decrease of Young's modulus and tensile strength of SPEEK are attributable to the plasticizing effect of liquid water due to its high dielectric constant, the behavior of Nafion^{TM} in the buffer is more surprising. One notices a very strong reduction of the elongation at break and a significant increase of Young's modulus. The enhanced stiffness (and reduced ductility) of Nafion^{TM} is corroborated by the handling experience after buffer immersion.

Analyzing the mechanical test results (Table 3 and Figure 3), the behavior of SPEEK moves from a rigid polymer to a plastic one: The Young's modulus decreased and the elongation at break increased remarkably. This is certainly due to the plasticizing effect of the water inside the membrane, weakening the interactions between the macromolecular chains and/or functionalized groups [40,42]. In the case of Nafion^{TM}, the membrane evolves from a plastic to a rigid behavior after immersion inside the buffer solution. As already demonstrated [41], Nafion^{TM} exchanged with different cations shows a distinct shift to higher temperature of the glass transition, indicating an enhancement of the stiffness. Similar findings were reported in the literature: ion exchange in hydrated Nafion^{TM} samples increases the Young's modulus of the membranes in increasing order of ionic radius [43]. An increase in Young's modulus means that the material becomes stiffer and a larger force is necessary to cause elastic deformation.

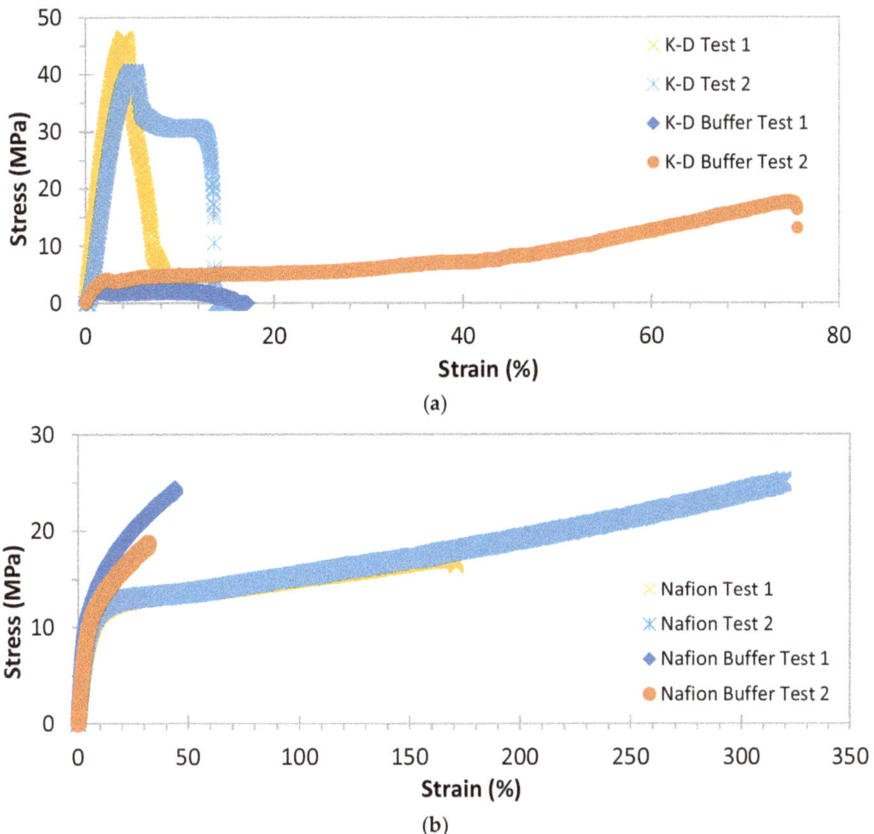

Figure 3. Typical stress-strain curves at relative humidity (RH) = 40% ± 10% and after immersion in buffer solution of (**a**) K-D and (**b**) Nafion 212 membrane.

Table 3. Mass uptake (MU), ionic conductivity, σ, and mechanical properties after conditioning in humid air or buffer solution at room temperature (RT). * in fully hydrated conditions.

Membrane	Conditions	MU (%)	σ (mS/cm)	Young's Modulus (MPa)	Yield Stress (MPa)	Ultimate Strength (MPa)	Elongation at Break (%)
Nafion™ 212	RH = 40% ± 10%	19.0	15.4 *	203 ± 26	5 ± 1	21 ± 6	245 ± 105
	Buffer	9.0	1.9	352 ± 88	5 ± 2	22 ± 4	38 ± 4
K-D	RH = 40% ± 10%	diss	diss	1423 ± 209	25 ± 8	44 ± 4	9 ± 6
	Buffer	74.3	15.8	295 ± 55	3 ± 2	10 ± 8	45 ± 30

4. Discussion

The difference in mechanical and solubility properties of cast and thermally treated (annealed and commercial) Nafion™ films is usually ascribed to the thermal reorganization of the cold cast micellar structure with sulfonate groups on the outside of the micelle to an inverted micellar structure with the sulfonate groups on the inside [44].

If we compare membranes cast from 100% SPEEK in different solvents and with different thermal treatment times (Figure 4), we can observe that the thermal treatment time (18 h or 15 min) does not affect the MU and conductivity and that membranes cast from DMSO and ethanol solutions have comparable properties. However, a very large enhancement of solution uptake and conductivity is observed for ethanol-water mixtures. This can be due to the different nanostructure in the membranes related to the high dielectric constant of water: the presence of water allows the formation of ionic clusters between sulfonic acid groups that are responsible for the swelling and the ionic conductivity [45]. The higher conductivity can also be explained by the large ionic mobility in the presence of a large amount of water (see discussion below).

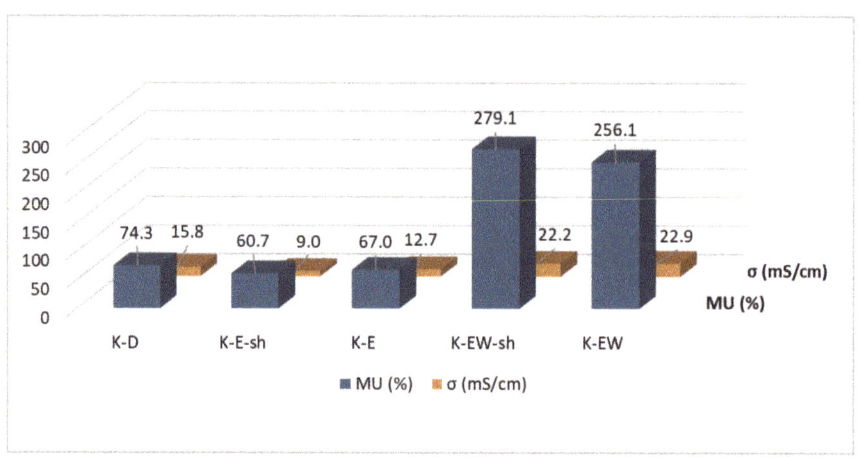

Figure 4. Comparison of SPEEK membranes realized with different solvents and casting procedures.

The thermal treatment of 18 h has a clear influence on the elastic modulus, which is much lower in comparison with the samples heated for 15 min only. The initially very stiff membranes deform easier. One can also observe that samples made from pure DMSO show the highest Young's modulus, and this high stiffness was reported before [11,37,46]. The presence of water in the casting solvent generally further enhances the Young's modulus. Membranes cast from pure ethanol instead show a quite low stiffness and, in the case of SPEEK, a low density. This indicates a low packing density of the macromolecular chains and low nanophase separation in accordance with the low conductivity.

Comparing membranes realized in 100% DMSO in Table 2, we can observe that the 100% SPPSU membrane dissolves in the buffer solution. The addition of 50% of SPEEK stabilizes the membrane but the apparent solution uptake reveals a partial dissolution phenomenon, attributable to some SPPSU loss in the buffer solution. The two membranes can be further stabilized by a crosslinking treatment. By adding 70% of SPEEK, the membrane does not dissolve, and the conductivity is high due to the effect of a large solution uptake.

Figure 5 shows membranes realized in 10% DMSO and 90% water composed of SPPSU or blended with SPEEK. We can observe, as already mentioned for the SPEEK membranes, that the addition of water to the casting solvent negatively affects the hydrolytic stability of the membranes. The 30/70 membrane (KU30-DW) has a higher MU than a membrane cast in pure DMSO and can be stabilized only with a crosslinking treatment. XL membranes present a good ionic conductivity, and their properties are similar to the reference SPEEK membrane cast from DMSO (K-D).

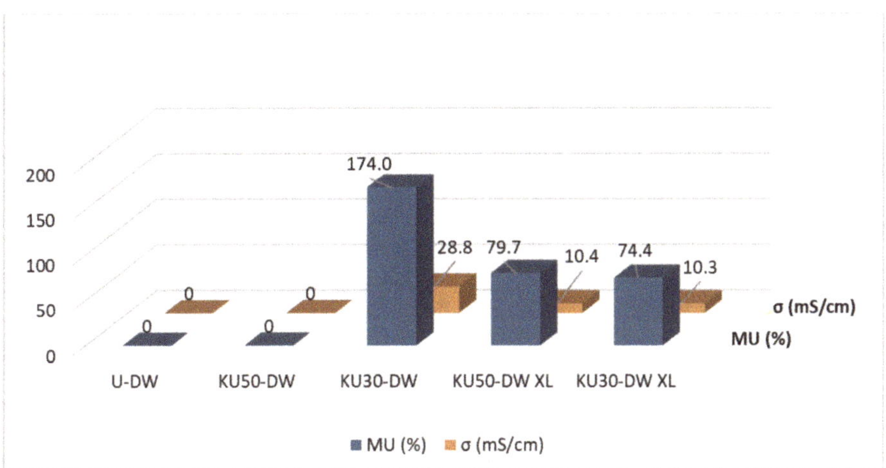

Figure 5. Comparison of SPEEK/SPPSU blend membranes cast from 10% DMSO and 90% water and crosslinked.

Membranes with 100% SPPSU cast from ethanol with two different thermal treatment times (U-E-sh and U-E in Table 2) are not stable, and the solvent and the treatment time do not modify the stability. These membranes need to be further crosslinked to achieve the right properties.

Figure 6 shows the ionic conductivity as a function of the mass uptake (MU). The apparent maximum at intermediate values of MU was reported before in other ionomers [47]. It is related to antagonistic effects of a cation concentration increase that simultaneously reduces the cation mobility so that an optimal value is obtained for an intermediate value of concentration.

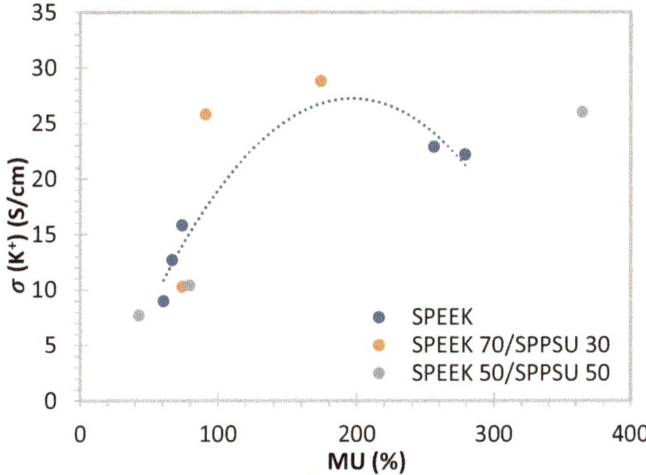

Figure 6. K^+ ion conductivity as a function of the mass uptake. The dotted line is a guide for the eye only.

We can further utilize all the data from the mass uptake and conductivity measurement to calculate an effective cation mobility (K^+), assuming that the ion exchange groups have been fully substituted by potassium. Furthermore, we assume that no supplementary adsorbed ions are present, because the exchange was made in a 0.05 M buffer so that the driving force for excess ion adsorption is low. The cation concentration expressed in mol/L in the membrane is then defined by the *IEC* and the *MU*. The density, ρ, of the electrolytic solution is taken as 1 g/cm^3 [48,49].

$$c(K^+) = \frac{IEC \times \rho}{MU} \tag{3}$$

The effective mobility $u(K^+)$ (in cm^2 V^{-1} s^{-1}) is defined according to Equation (4) by the measured ionic conductivity $\sigma(K^+)$ (expressed in mS/cm, Table 2) and the calculated ion concentration $c(K^+)$ (in mol/L):

$$u(K^+) = \frac{\sigma(K^+)}{F\,a \cdot c(K^+)} \tag{4}$$

The calculated cation mobility data are reported in Figure 7.

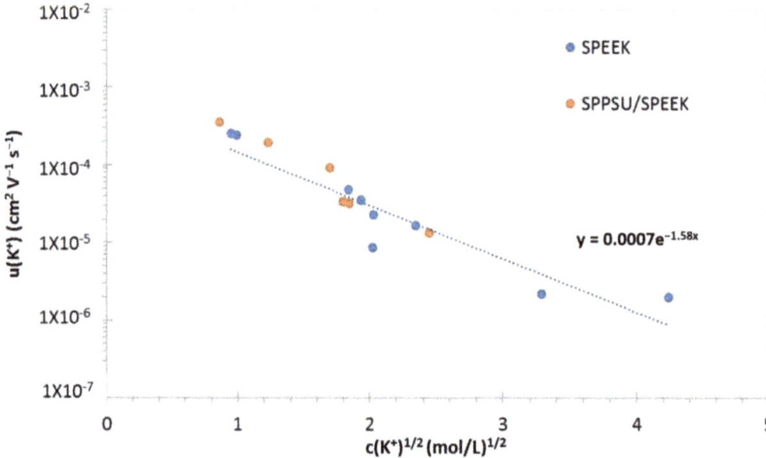

Figure 7. K^+ ion mobility in various proton exchange membranes based on SPEEK and SPPSU/SPEEK blends.

The exponential dependence of the cation mobility as a function of the square root of cation concentration has been previously related to the conditions of ionic motion in nanometric ion conduction channels, where the mobile ions migrate with immobile counterions grafted on the channel walls [50–52]. The extrapolation to $c(i) = 0$ gives the cation mobility at infinite dilution ($u(K^+) = 7 \times 10^{-4}$ cm^2 V^{-1} s^{-1}) that can be compared with literature data of K^+ mobility in aqueous solution ($u(K^+) = 7.6 \times 10^{-4}$ cm^2 V^{-1} s^{-1}) [53,54].

5. Conclusions

This work explored the possibility to adapt the hydrolytic behavior, the conductivity and the mechanical properties of ion-conducting membranes based on sulfonated aromatic polymers such as SPEEK and SPPSU by changing the casting solvent and procedure, by blending two different polymers and by a crosslinking treatment. The results of the different characterizations were discussed to find out the optimal compromise for the utilization of these polymers as membranes in EFC containing a buffer solution. The solvent and the casting procedure influenced the properties of the membranes: the use of ethanol reduced the swelling, but at the same time also the Young's modulus. A shorter casting time resulted in stiffer membranes but had no effect on MU and conductivity. The crosslinking stabilized the membranes that showed a better hydrolytic stability, even if with a little decrease of conductivity. The addition of SPPSU to SPEEK membranes improved the ionic conductivity; however, the ratio between the two polymers has to be carefully tuned in order to keep the hydrolytic stability.

The comparison between a DMSO-cast SPEEK membrane and Nafion^{TM} shows how the behavior of the two polymers before and after the immersion in the buffer solution is different. The SPEEK membrane exhibited a higher MU and conductivity in comparison to Nafion^{TM}, and this difference can be ascribed to the solution uptake and the quite high cation mobility in this SAP. The mechanical properties change differently in the two polymers: SPEEK moves from a rigid to a plastic behavior while Nafion^{TM} does exactly the contrary.

The analysis of all data demonstrated that an optimal value of conductivity can be obtained at intermediate values of mass uptake and that the extrapolation to infinite dilution of the mobility of conducting potassium ions is in good agreement with literature data.

This study opens interesting perspectives for the utilization of SAP polymers as membranes in enzymatic and Bio-FC, exploring the possibility of less hazardous casting solvents, while maintaining appropriate conductivity and hydrolytic stability of membranes in buffer solutions.

Author Contributions: Conceptualization, L.P.; Data curation, L.P., B.Z. and E.S.; Formal analysis, P.K.; Methodology, L.P.; Project administration, L.P.; Supervision, E.S., R.N., M.L.D.V. and P.K.; Validation, M.L.D.V. and P.K.; Writing—original draft, L.P.; Writing—review & editing, E.S., R.N., M.L.D.V. and P.K. All authors have read and agreed to the published version of the manuscript.

Funding: The project leading to this publication has received funding from the Excellence Initiative of Aix-Marseille Université-A*MIDEX, a French "Investissements d'Avenir" programme", AMX-18-MED-002 ENZIM-FC.

Institutional Review Board Statement: Not applicable.

Informed Consent Statement: Not applicable.

Data Availability Statement: Not applicable.

Conflicts of Interest: The authors declare no conflict of interest.

References

1. Zhang, W.; Shi, Z.; Zhang, F.; Liu, X.; Jin, J.; Jiang, L. Superhydrophobic and superoleophilic pvdf membranes for effective separation of water-in-oil emulsions with high flux. *Adv. Mater.* **2013**, *25*, 2071–2076. [CrossRef] [PubMed]
2. Skulimowska, A.; Dupont, M.; Zatoń, M.; Sunde, S.; Merlo, L.; Jones, D.; Rozière, J. Proton exchange membrane water electrolysis with short-side-chain Aquivion® membrane and IrO_2 anode catalyst. *Int. J. Hydrog. Energy* **2014**, *39*, 6307–6316. [CrossRef]
3. Marini, S.; Salvi, P.; Nelli, P.; Pesenti, R.; Villa, M.; Berrettoni, M.; Zangari, G.; Kiros, Y. Advanced alkaline water electrolysis. *Electrochim. Acta* **2012**, *82*, 384–391. [CrossRef]
4. Doan, T.N.L.; Hoang, T.K.A.; Chen, P. Recent development of polymer membranes as separators for all-vanadium redox flow batteries. *RSC Adv.* **2015**, *5*, 72805–72815. [CrossRef]
5. Kreuer, K.D. Ion conducting membranes for fuel cells and other electrochemical devices. *Chem. Mater.* **2014**, *26*, 361–380. [CrossRef]
6. Hess, M.; Jones, R.G.; Kahovec, J.; Kitayama, T.; Kratochvíl, P.; Kubisa, P.; Mormann, W.; Stepto, R.F.T.; Tabak, D.; Vohlídal, J.; et al. Terminology of polymers containing ionizable or ionic groups and of polymers containing ions (IUPAC Recommendations 2006). *Pure Appl. Chem.* **2006**, *78*, 2067–2074. [CrossRef]
7. Zahn, R.; Vörös, J.; Zambelli, T. Swelling of electrochemically active polyelectrolyte multilayers. *Curr. Opin. Colloid Interface Sci.* **2010**, *15*, 427–434. [CrossRef]
8. Kreuer, K.-D. The role of internal pressure for the hydration and transport properties of ionomers and polyelectrolytes. *Solid State Ion.* **2013**, *252*, 93–101. [CrossRef]
9. Knauth, P.; Sgreccia, E.; Di Vona, M. Chemomechanics of acidic ionomers: Hydration isotherms and physical model. *J. Power Sources* **2014**, *267*, 692–699. [CrossRef]
10. Kim, J.-D.; Donnadio, A.; Jun, M.-S.; Di Vona, M.L. Crosslinked SPES-SPPSU membranes for high temperature PEMFCs. *Int. J. Hydrog. Energy* **2013**, *38*, 1517–1523. [CrossRef]
11. Hou, H.; Di Vona, M.L.; Knauth, P. Building bridges: Crosslinking of sulfonated aromatic polymers—A review. *J. Membr. Sci.* **2012**, *423–424*, 113–127. [CrossRef]
12. De Poulpiquet, A.; Ranava, D.; Monsalve, K.; Giudici-Orticoni, M.-T.; Lojou, E. Biohydrogen for a new generation of H_2/O_2 Biofuel cells: A sustainable energy perspective. *Chemelectrochem* **2014**, *1*, 1724–1750. [CrossRef]

13. Mazurenko, I.; Monsalve, K.; Infossi, P.; Giudici-Orticoni, M.-T.; Topin, F.; Mano, N.; Lojou, E. Impact of substrate diffusion and enzyme distribution in 3D-porous electrodes: A combined electrochemical and modelling study of a thermostable H_2/O_2 enzymatic fuel cell. *Energy Environ. Sci.* **2017**, *10*, 1966–1982. [CrossRef]
14. Bullen, R.A.; Arnot, T.C.; Lakeman, J.B.; Walsh, F.C. Biofuel cells and their development. *Biosens. Bioelectron.* **2006**, *21*, 2015–2045. [CrossRef] [PubMed]
15. Squadrito, G.; Cristiani, P. Microbial and enzymatic fuel cells. In *Compendium of Hydrogen Energy*; Subramani, V., Bsile, A., Verziroglu, N., Eds.; Elsevier: Amsterdam, The Netherlands, 2016; pp. 147–173.
16. Rosenbaum, M.A.; Aulenta, F.; Villano, M.; Angenent, L.T. Cathodes as electron donors for microbial metabolism: Which extracellular electron transfer mechanisms are involved? *Bioresour. Technol.* **2011**, *102*, 324–333. [CrossRef]
17. Cristiani, P.; Franzetti, A.; Gandolfi, I.; Guerrini, E.; Bestetti, G. Bacterial DGGE fingerprints of biofilms on electrodes of membraneless microbial fuel cells. *Int. Biodeterior. Biodegrad.* **2013**, *84*, 211–219. [CrossRef]
18. Yahiro, A.T.; Lee, S.; Kimble, D.O. Bioelectrochemistry. *Biochim. Biophys. Acta Sect. Biophys. Subj.* **1964**, *88*, 375–383. [CrossRef]
19. Barton, S.C.; Gallaway, J.; Atanassov, P. Enzymatic biofuel cells for implantable and microscale devices. *Chem. Rev.* **2004**, *104*, 4867–4886. [CrossRef]
20. Naidoo, S.; Naidoo, Q.; Blottnitz, H.; Vaivars, G. Glucose oxidase as a biocatalytic enzyme-based bio-fuel cell using Nafion membrane limiting crossover. *IOP Conf. Ser. Mater. Sci. Eng.* **2013**, *49*, 012062. [CrossRef]
21. Moehlenbrock, M.J.; Minteer, S.D. Extended lifetime biofuel cells. *Chem. Soc. Rev.* **2008**, *37*, 1188–1196. [CrossRef]
22. Lalaoui, N.; De Poulpiquet, A.; Haddad, R.; Le Goff, A.; Holzinger, M.; Gounel, S.; Mermoux, M.; Infossi, P.; Mano, N.; Lojou, E.; et al. A membraneless air-breathing hydrogen biofuel cell based on direct wiring of thermostable enzymes on carbon nanotube electrodes. *Chem. Commun.* **2015**, *51*, 7447–7450. [CrossRef] [PubMed]
23. Mazurenko, I.; Wang, X.; De Poulpiquet, A.; Lojou, E. H2/O2enzymatic fuel cells: From proof-of-concept to powerful devices. *Sustain. Energy Fuels* **2017**, *1*, 1475–1501. [CrossRef]
24. Xiao, X.; Xia, H.-Q.; Wu, R.; Bai, L.; Yan, L.; Magner, E.; Cosnier, S.; Lojou, E.; Zhu, Z.; Liu, A. tackling the challenges of enzymatic (bio)fuel cells. *Chem. Rev.* **2019**, *119*, 9509–9558. [CrossRef] [PubMed]
25. Pasquini, L.; Wacrenier, O.; Di Vona, M.; Knauth, P. Hydration and ionic conductivity of model cation and anion-conducting ionomers in buffer solutions (Phosphate, Acetate, Citrate). *J. Phys. Chem. B* **2018**, *122*, 12009–12016. [CrossRef] [PubMed]
26. Choi, M.-J.; Chae, K.-J.; Ajayi, F.F.; Kim, K.-Y.; Yu, H.-W.; won Kim, C.; Kim, I.S. Effects of biofouling on ion transport through cation exchange membranes and microbial fuel cell performance. *Bioresour. Technol.* **2011**, *102*, 298–303. [CrossRef]
27. Xu, J.; Sheng, G.-P.; Luo, H.-W.; Li, W.-W.; Wang, L.-F.; Yu, H.-Q. Fouling of proton exchange membrane (PEM) deteriorates the performance of microbial fuel cell. *Water Res.* **2012**, *46*, 1817–1824. [CrossRef]
28. Rozendal, R.A.; Hamelers, H.V.M.; Buisman, C.J.N. Effects of membrane cation transport on pH and microbial fuel cell performance. *Environ. Sci. Technol.* **2006**, *40*, 5206–5211. [CrossRef]
29. Bahar, T.; Yazici, M.S. Performance assessment of a perfluorosulfonic acid-type membrane (i.e., Nafion™ 115) for an enzymatic fuel cell. *Electroanalysis* **2019**, *31*, 1656–1663. [CrossRef]
30. Rozendal, R.A.; Hamelers, H.V.; Molenkamp, R.J.; Buisman, C.J.N. Performance of single chamber biocatalyzed electrolysis with different types of ion exchange membranes. *Water Res.* **2007**, *41*, 1984–1994. [CrossRef]
31. Di Vona, M.L.; Luchetti, L.; Spera, G.P.; Sgreccia, E.; Knauth, P. Synthetic strategies for the preparation of proton-conducting hybrid polymers based on PEEK and PPSU for PEM fuel cells. *Comptes Rendus Chim.* **2008**, *11*, 1074–1081. [CrossRef]
32. Zaidi, S.M.J.; Mikhailenko, S.D.; Robertson, G.; Guiver, M.; Kaliaguine, S. Proton conducting composite membranes from polyether ether ketone and heteropolyacids for fuel cell applications. *J. Membr. Sci.* **2000**, *173*, 17–34. [CrossRef]
33. Di Vona, M.L.; Sgreccia, E.; Licoccia, S.; Alberti, G.; Tortet, L.; Knauth, P. Analysis of temperature-promoted and solvent-assisted cross-linking in sulfonated poly(ether ether ketone) (SPEEK) proton-conducting membranes. *J. Phys. Chem. B* **2009**, *113*, 7505–7512. [CrossRef] [PubMed]
34. Di Vona, M.L.; Alberti, G.; Sgreccia, E.; Casciola, M.; Knauth, P. High performance sulfonated aromatic ionomers by solvothermal macromolecular synthesis. *Int. J. Hydrog. Energy* **2012**, *37*, 8672–8680. [CrossRef]
35. Maranesi, B.; Hou, H.; Polini, R.; Sgreccia, E.; Alberti, G.; Narducci, R.; Knauth, P.; Di Vona, M.L. Cross-linking of sulfonated poly(ether ether ketone) by thermal treatment: How does the reaction occur? *Fuel Cells* **2013**, *13*, 107–117. [CrossRef]
36. Hitaishi, V.P.; Mazurenko, I.; Harb, M.; Clément, R.; Taris, M.; Castano, S.; Duché, D.; Lecomte, S.; Ilbert, M.; De Poulpiquet, A.; et al. Electrostatic-driven activity, loading, dynamics, and stability of a redox enzyme on functionalized-gold electrodes for bioelectrocatalysis. *ACS Catal.* **2018**, *8*, 12004–12014. [CrossRef]
37. Barbieri, G.; Brunetti, A.; Di Vona, M.L.; Sgreccia, E.; Knauth, P.; Hou, H.; Hempelmann, R.; Arena, F.; Beretta, L.D.; Bauer, B.; et al. LoLiPEM: Long life proton exchange membrane fuel cells. *Int. J. Hydrog. Energy* **2016**, *41*, 1921–1934. [CrossRef]
38. Knauth, P.; Pasquini, L.; Di Vona, M.L. Comparative study of the cation permeability of protonic, anionic and ampholytic membranes. *Solid State Ion.* **2017**, *300*, 97–105. [CrossRef]
39. Knauth, P.; Pasquini, L.; Maranesi, B.; Pelzer, K.; Polini, R.; Di Vona, M.L. Proton mobility in sulfonated polyetheretherketone (speEK): Influence of thermal crosslinking and annealing. *Fuel Cells* **2013**, *13*, 79–85. [CrossRef]
40. Narducci, R.; Di Vona, M.L.; Knauth, P. Cation-conducting ionomers made by ion exchange of sulfonated poly-ether-ether-ketone: Hydration, mechanical and thermal properties and ionic conductivity. *J. Membr. Sci.* **2014**, *465*, 185–192. [CrossRef]

41. Chailan, J.-F.; Khadhraoui, M.; Knauth, P. Mechanical and dynamic mechanical analysis of proton-conducting polymers. In *Solid State Proton Conductors*; Di Vona, M.L., Knauth, P., Eds.; Wiley: Chichester, UK, 2012; pp. 222–225.
42. Kim, Y.S.; Dong, L.; Hickner, M.A.; Pivovar, B.S.; McGrath, J.E. Processing induced morphological development in hydrated sulfonated poly(arylene ether sulfone) copolymer membranes. *Polymer* **2003**, *44*, 5729–5736. [CrossRef]
43. Kundu, S.; Simon, L.C.; Fowler, M.; Grot, S. Mechanical properties of Nafion™ electrolyte membranes under hydrated conditions. *Polymer* **2005**, *46*, 11707–11715. [CrossRef]
44. Zook, L.A.; Leddy, J. Density and solubility of nafion: Recast, annealed, and commercial films. *Anal. Chem.* **1996**, *68*, 3793–3796. [CrossRef] [PubMed]
45. Koziara, B.T.; Akkilic, N.; Nijmeijer, D.C.; Benes, N.E. The effects of water on the morphology and the swelling behavior of sulfonated poly(ether ether ketone) films. *J. Mater. Sci.* **2016**, *51*, 1074–1082. [CrossRef]
46. Di Vona, M.L.; Sgreccia, E.; Tamilvanan, M.; Khadhraoui, M.; Chassigneux, C.; Knauth, P. High ionic exchange capacity polyphenylsulfone (SPPSU) and polyethersulfone (SPES) cross-linked by annealing treatment: Thermal stability, hydration level and mechanical properties. *J. Membr. Sci.* **2010**, *354*, 134–141. [CrossRef]
47. Di Vona, M.L.; Pasquini, L.; Narducci, R.; Pelzer, K.; Donnadio, A.; Casciola, M.; Knauth, P. Cross-linked sulfonated aromatic ionomers via SO2 bridges: Conductivity properties. *J. Power Sources* **2013**, *243*, 488–493. [CrossRef]
48. Knauth, P.; Sgreccia, E.; Donnadio, A.; Casciola, M.; Di Vona, M.L. Water activity coefficient and proton mobility in hydrated acidic polymers. *J. Electrochem. Soc.* **2011**, *158*, B159–B165. [CrossRef]
49. Knauth, P.; Di Vona, M. Sulfonated aromatic ionomers: Analysis of proton conductivity and proton mobility. *Solid State Ion.* **2012**, *225*, 255–259. [CrossRef]
50. Mauritz, K.A.; Moore, R.B. State of understanding of nafion. *Chem. Rev.* **2004**, *104*, 4535–4586. [CrossRef]
51. Gierke, T.D.; Munn, G.E.; Wilson, F.C. The morphology in nafion perfluorinated membrane products, as determined by wide- and small-angle x-ray studies. *J. Polym. Sci. Polym. Phys. Ed.* **1981**, *19*, 1687–1704. [CrossRef]
52. Yeager, H.L. Cation and water diffusion in nafion ion exchange membranes: Influence of polymer structure. *J. Electrochem. Soc.* **1981**, *128*, 1880. [CrossRef]
53. Hamann, C.H.; Hamnett, A.; Vielstich, W. *Electrochemistry*, 2nd ed.; Wiley: Weinheim, Germany, 2007.
54. Jaffrezic-Renault, N.; Dzyadevych, S.V. Conductometric microbiosensors for environmental monitoring. *Sensors* **2008**, *8*, 2569–2588. [CrossRef] [PubMed]

Article

Enhanced Hydroxide Conductivity and Dimensional Stability with Blended Membranes Containing Hyperbranched PAES/Linear PPO as Anion Exchange Membranes

Sang Hee Kim [1], Kyu Ha Lee [2], Ji Young Chu [2], Ae Rhan Kim [1,2] and Dong Jin Yoo [1,2,*]

1 Department of Energy Storage/Conversion Engineering, Hydrogen and Fuel Cell Research Center, Jeonbuk National University, Jeonju 54896, Jeollabuk-do, Korea; tkdgml3257@naver.com (S.H.K.); canutech@hanmail.net (A.R.K.)

2 Department of Life Sciences, Jeonbuk National University, Jeonju 54896, Jeollabuk-do, Korea; ebbuneg@hanmail.net (K.H.L.); carumiss@naver.com (J.Y.C.)

* Correspondence: djyoo@jbnu.ac.kr; Tel.: +82-63-270-3608

Received: 1 December 2020; Accepted: 15 December 2020; Published: 16 December 2020

Abstract: A series of novel blended anion exchange membranes (AEMs) were prepared with hyperbranched brominated poly(arylene ether sulfone) (Br-HB-PAES) and linear chloromethylated poly(phenylene oxide) (CM-PPO). The as-prepared blended membranes were fabricated with different weight ratios of Br-HB-PAES to CM-PPO, and the quaternization reaction for introducing the ionic functional group was performed by triethylamine. The Q-PAES/PPO-XY (quaternized-PAES/PPO-XY) blended membranes promoted the ion channel formation as the strong hydrogen bonds interconnecting the two polymers were maintained, and showed an improved hydroxide conductivity with excellent thermal behavior. In particular, the Q-PAES/PPO-55 membrane showed a very high hydroxide ion conductivity (90.9 mS cm^{-1}) compared to the pristine Q-HB-PAES membrane (32.8 mS cm^{-1}), a result supported by the morphology of the membrane as determined by the AFM analysis. In addition, the rigid hyperbranched structure showed a suppressed swelling ratio of 17.9–24.9% despite an excessive water uptake of 33.2–50.3% at 90 °C, and demonstrated a remarkable alkaline stability under 2.0 M KOH conditions over 1000 h.

Keywords: anion exchange membrane; hyperbranched polymer; alkaline fuel cell; blended membranes; anion conductivity; dimensional stability

---

1. Introduction

Fuel cells (FCs) are a promising energy conversion device, that can directly convert chemical energy into electrical energy [1–3]. Among the various existing FCs, cation exchange membrane fuel cells (CEMFCs) have received particular attention due to their many advantages, such as high power density and superior efficiency [4,5]. Despite these advantages, the commercialization of the CEMFC has been difficult due to the high system costs associated with the use of precious metal platinum catalysts and fluorinated membranes [6]. On the other hand, anion exchange membrane fuel cells (AEMFCs) are capable of reducing system costs by using less valuable metal catalysts (Co and Ni) [7] and have been studied recently due to their advantages, including fast oxygen reduction reactions (ORR) at the cathode, low fuel crossover, and easy water management. To advance AEMFCs, the development of an anion exchange membrane (AEM) with high ionic conductivity and physicochemical stability is essential [8]. However, the AEMs that have been studied thus far yield an insufficient electrochemical performance due to the inherently lower mobility of OH^- ions compared to H^+ ions and unstable dimensional stability due to the excessive swelling of the AEM [9–12]. Therefore, many researchers have studied AEM with high ion exchange capacity (IEC) value as a method of improving ionic conductivity. However, a high IEC value causes excessive dimensional change in the membrane along with a high water uptake, which leads to separation of the membrane electrode assembly (MEA) during the manufacture and deterioration of fuel cell durability from the wet-dry repetition [13,14]. As mentioned above, various studies have achieved excellent dimensional stability along with high IEC values such as blended membranes [15–18]; inorganic/organic composite membranes [6,19,20]; and the backbone design of comb-shaped [21–23], branching [24–26], and cross-linking [27,28].

Novel hyperbranched anion exchange membranes (HBAEMs) are promising backbone structures for AEMs that have been reported to possess improved dimensional stability, alkaline stability, and clear micro-phase separation of hydrophilic/hydrophobic domains, which result in excellent ionic conductivity and high solubility enabling facile membrane fabrication and reprocessing compared to other polymers [29–32]. Ge et al. [33] reported hyperbranched poly-4-(chloromethyl) styrene (HB-PVBC) AEMs with superior dimensional stability and an ionic conductivity of 123.01 mS cm^{-1} at 80 °C, which results from the well-separated hydrophilic/hydrophobic microphases and the low swelling ratio of HB-PVBC due to its hyperbranched structure.

In addition, a promising strategy for the development of AEMs with excellent ionic conductivity, mechanical properties, and thermal stability is to blend polymers to achieve a balance of physicochemical stability and electrochemical performance [34]. For example, Kim et al. [35,36] reported that miscible blended membranes comprised of sulfonated-fluorinated, hydrophilic-hydrophobic copolymer and sulfonated poly(ether ketone) demonstrated good phase-separated morphology and superior ionic conductivity.

Recently, the aromatic polymer, poly(phenylene oxide) (PPO) has been used in many applications due to its good thermal stability, excellent mechanical strength, improved conductivity via the introduction of functional groups, and facile synthesis [37]. Xu et al. [38] reported that the development of a series of PPO-based AEMs with ethylene oxide spacers demonstrated a very good peak power density (437 mW cm^{-2}).

In accordance with these studies, we designed a blended membrane of hyperbranched poly(arylene ether sulfone) (HB-PAES) and the linear polymer poly(phenylene oxide) (PPO) as a novel strategy. This strategy aims to compensate for each polymer's drawbacks [39]. In this study, bromination and chloromethylation reactions were performed to introduce quaternary functional groups into HBPAES based on the hyperbranched structure and PPO polymer, respectively, and the structure of the prepared polymers was confirmed through ^{1}H NMR, FT-IR, and GPC analyses. Subsequently, a series of blended polymers were fabricated by adjusting the mixing ratio of HBPAES to PPO (3:7, 4:6, 5:5, 6:4, and 7:3) to investigate, the physicochemical stability and electrochemical properties affected by the mixing ratio.

2. Materials and Methods

2.1. Materials

1,1,1-Tris(4-hydroxyphenyl)ethane (THPE) and 2,2-bis(4-hydroxy-3-methylphenyl)propane (BHMP) were purchased from Tokyo Chemical Industries (Tokyo, Japan). Bis-(4-chlorophentyl)sulfone (BPS), N-bromosuccinimide (NBS), anhydrous toluene, anhydrous *N,N*-dimethylacetamide (DMAc), poly(2,6-dimethyl-1,4-phenylene oxide) (PPO), and triethylamine (TEA) were obtained from Sigma-Aldrich (Seoul, Korea). Potassium carbonate (K_2CO_3), benzoyl peroxide (BPO), 1,1,2,2-tetrachloroethane (TCE), ethanol, methanol, and acetone were purchased from Daejung Chemicals and Metals company (Seoul, Korea).

2.2. Synthesis of the Hyperbranched Hydrophobic Oligomer

The hyperbranched Cl-terminated poly(arylene ether sulfone) hydrophobic oligomer (HB-PAES-Cl) was synthesized via the aromatic nucleophilic substitution reaction (Scheme 1a). The polymerization of HB-PAES-Cl was carried out in THPE (1.25 g, 4.08 mmol), BPS (2.58 g, 8.98 mmol), K_2CO_3 (1.69 g, 12.24 mmol), DMAc (30.0 mL), and toluene (15.0 mL) in a round-bottomed flask equipped with a Dean-Stark trap and a reflux condenser under nitrogen gas. The mixture was heated to 150 °C under stirring, and its temperature was maintained at 150 °C for 5 h to remove azeotropes (toluene and water). Subsequently, polymerization was carried out at 180 °C for 24 h. Afterward, the mixture was cooled at room temperature, and the viscous solution was poured into a mixture of methanol, acetone, and distilled water (DI water) (*v*/*v*/*v*, 7/1/1). The precipitated powder was collected by filtration, and the white fiber product was dried in a vacuum oven at 70 °C for 24 h (yield: 92%). The chemical structure of the as-synthesized HB-PAES-Cl was characterized using ^{1}H NMR, as shown in Figure S1a (600 MHz, DMSO-d_6: 2.17–2.00 ppm (H_n), 7.16–6.84 ppm (H_h, $H_{h'}$, H_l, H_m), 7.69–7.54 ppm (H_k), 7.97–7.75 ppm ($H_{i'}$, H_j, $H_{j'}$), and GPC (M_n = 7 kDa, M_w = 22 kDa).

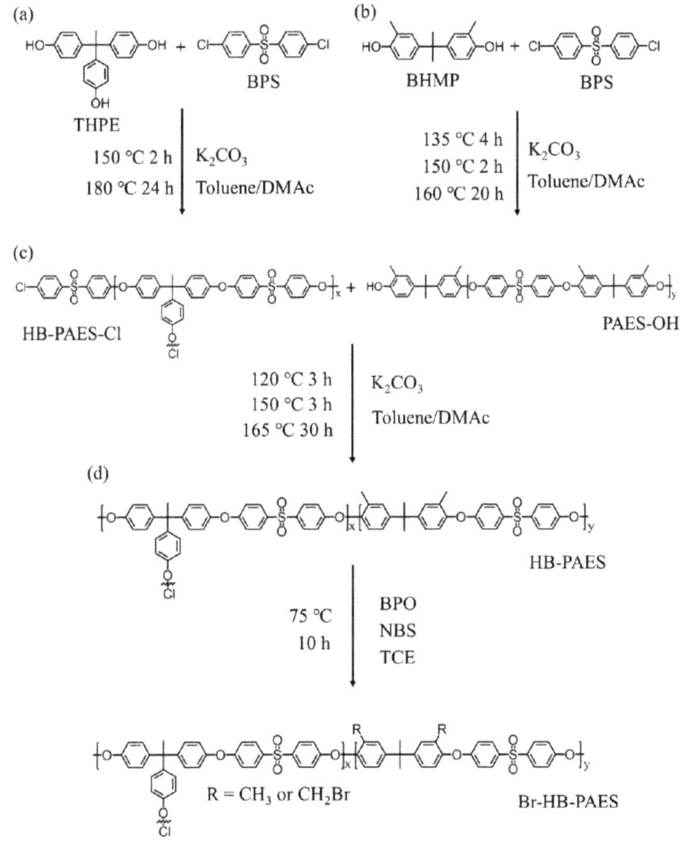

Scheme 1. Synthesis of (**a**) HB-PAES-Cl, (**b**) PAES-OH, (**c**) HB-PAES, and (**d**) Br-HB-PAES.

2.3. Synthesis of the Hydrophilic Oligomer

The OH-terminated poly(arylene ether sulfone) hydrophilic oligomer (PAES-OH) was synthesized via the aromatic nucleophilic substitution reaction, in a process similar to that used for HB-PAES-Cl [40]. As shown in Scheme 1b, the polymerization of PAES-OH was performed with BHMP (1.96 g, 7.66 mmol), BPS (2.00 g, 6.96 mmol), K_2CO_3 (2.12 g, 15.32 mmol), DMAc (20.0 mL), and toluene (15.0 mL) in a round-bottomed flask equipped with a Dean-Stark trap and a reflux condenser under nitrogen gas. The reaction mixture was heated to 135 °C under stirring, and its temperature was maintained at

150 °C for 2 h to remove azeotropes (toluene and water). Afterward, polymerization was carried out at 160 °C for 20 h. Subsequently, the mixture was cooled at room temperature, and the viscous solution was poured into a mixture of methanol, acetone, and DI water (*v*/*v*/*v*, 7/1/1). Finally, the precipitated powder was collected by filtration, and the light brown product was dried in a vacuum oven at 70 °C for 15 h (yield: 94%). The chemical structure of the as-synthesized HTHP-OH was characterized using ^{1}H NMR, as shown in Figure S1b (600 MHz, DMSO-d_6: 1.66–1.39 ppm (H_g), 2.07–1.81 ppm (H_f), 7.07–6.63 ppm (H_a, H_b, H_d), 7.26–7.12 ppm (H_c), 7.87–7.69 ppm (H_e), and GPC (M_n = 10 kDa, M_w = 22 kDa).

2.4. Synthesis of Hyperbranched Poly(Arylene Ether Sulfone) Block Copolymer

The hyperbranched poly(arylene ether sulfone) (HB-PAES) was prepared via the aromatic nucleophilic substitution reaction (Scheme 1c). The polymerization of HB-PAES was conducted with HB-PAES-Cl (2.00 g, 0.19 mmol), PAES-OH (2.14 g, 0.19 mmol), K_2CO_3 (0.05 g, 0.38 mmol), DMAc (25.0 mL), and toluene (12.0 mL) in a round-bottomed flask equipped with a Dean-Stark trap and reflux condenser under nitrogen flow. The reaction mixture was heated at 120 °C under stirring, and the mixture temperature was maintained at 150 °C for 3 h to remove azeotrope (toluene and water). Afterward, the polymerization was carried out at 165 °C for 30 h. Subsequently, the mixture was cooled to room temperature, and the viscous solution was poured into a mixture of methanol, acetone, and DI water (*v*/*v*/*v*, 7/1/1). Finally, the precipitated powder was collected by filtration, and the light brown product was dried in a vacuum oven at 70 °C for 15 h (yield: 90%). The chemical structure of the as-synthesized HB-PAES was characterized using ^{1}H NMR, as shown in Figure S1c (600 MHz, DMSO-d_6: 1.65–1.46 ppm (H_g), 2.15–1.80 ppm (H_f, H_n), 7.23–6.70 ppm (H_a, H_b, H_d, H_h, $H_{h'}$, H_l, H_m), 7.26–7.25 ppm (H_c), 7.55–7.54 ppm (H_k), 7.95–7.55 ppm (H_e, H_i, $H_{i'}$, H_j), and GPC (M_n = 97 kDa, M_w = 153 kDa).

2.5. Synthesis of Brominated HB-PAES (Br-HB-PAES)

The brominated hyperbranched poly(arylene ether sulfone) (Br-HB-PAES) was synthesized via the Friedel-Crafts alkylation (Scheme 1d) [41–43]. The bromination reaction of Br-HB-PAES was performed with HB-PAES (3.00 g, 0.14 mmol), TCE (30.0 mL), and NBS (4.49 g, 6.30 mmol) as the bromination reagent, with BPO (0.61 g, 0.63 mmol) as the initiator in a two-neck round bottom flask equipped with a reflux condenser under nitrogen flow. The reaction solution was heated at 40 °C under stirring, and the bromination reaction was carried out at 75 °C for 10 h. Afterward, the mixture was cooled at room temperature, and the yellow product was poured into methanol (700 mL). Finally, the precipitated powder was collected by filtration, and the yellowish product was dried in a vacuum oven at 70 °C for 15 h (yield: 95%). The chemical structure of the as-synthesized Br-HB-PAES was characterized using ^{1}H NMR (600 MHz, DMSO-d_6: 1.65–1.46 ppm, 2.15–1.80 ppm, 4.60–4.49 ppm, 7.23–6.70 ppm, 7.26–7.25 ppm, 7.55–7.54 ppm, 7.95–7.55 ppm).

2.6. Synthesis of Chloromethylated PPO (CM-PPO)

As shown in Scheme S1, for chloromethylation, the CM-PPO copolymer was synthesized via the Friedel-Crafts alkylation [6,41]. The PPO polymer (3.00 g, 0.10 mmol) was dissolved in TCE (15.0 mL) at 30 °C in a two-neck round-bottomed flask equipped with a reflux condenser under nitrogen. Then, chloromethyl ethyl ether (CMME) (7.0 mL) and $ZnCl_2$ (0.20 g) as a catalyst dispersed in tetrahydrofuran (THF) (2.0 mL) were added drop-wise to the reaction mixture. Subsequently, the mixture was reacted at 50 °C for 7 days. After cooling, the reaction mixture of chloromethylated PPO (CM-PPO) was precipitated in methanol. Finally, the precipitated powder was collected by filtration, and the white powder was dried in an oven at 60 °C for 24 h. The degree of chloromethylation (DC) was confirmed by ^{1}H NMR (600 MHz, DMSO-d_6: 2.27–1.89 ppm, 4.99–4.61 ppm, 6.14–5.99 ppm, 6.55–6.44 ppm).

2.7. Fabrication of the BC-PAES/PPO-XY Blended Membranes

The BC-PAES/PPO-XY (where X and Y represent the weight ratios of Br-HB-PAES and CM-PPO, respectively) blended membranes were prepared by the direct solution casting using different weight ratios of Br-HB-PAES to CM-PPO, as shown in Table 1. For example, the BC-PAES/PPO-55 blended membrane was fabricated with Br-HB-PAES (0.25 g), CM-PPO (0.25 g), and TCE (12.0 mL), and the solution was stirred at 50 °C for 10 h. The solution was cast onto a clean flat glass plate, and dried in a vacuum oven at 70 °C for 13 h. The BC-PAES/PPO-55 blended membrane was peeled off the glass plate using DI water. The fabricated blended membranes were flexible, transparent, and had an average thickness of 30–50 μm.

Table 1. Different weight ratios of BC-PAES/PPO-XY blended membranes.

Samples	Br-HB-PAES (%)	CM-PPO (%)
BC-PAES/PPO-37	30	70
BC-PAES/PPO-46	40	60
BC-PAES/PPO-55	50	50
BC-PAES/PPO-64	60	40
BC-PAES/PPO-73	70	30

2.8. Quaternization of BC-PAES/PPO-XY Membranes

Quaternized BC-PAES/PPO-XY (Q-PAES/PPO-XY) blended membranes were prepared using triethylamine (TEA) as a tertiary amine for the quaternization reaction. Briefly, the as-prepared membranes were immersed in a 25 wt% triethylamine solution at 50 °C for 12 h to exchange Cl or Br groups for quaternary ammonium groups. Afterward, the quaternized samples were soaked in a 1.0 M KOH solution for 48 h to convert Br^- or Cl^- forms to OH^- forms, and the quaternized membranes were kept in DI water.

3. Characterizations

The proton nuclear magnetic resonance (^{1}H NMR) spectroscopy was utilized to analyze the chemical structure of the polymers using the JNM-ECA 600 instruments (Bruker installed at the Center for University Wide Research Facilities (CURF) in Jeonbuk National University (JBNU), Jeonju, Korea. The number average molecular weight, weight average molecular weight, and polydispersity index (PDI) of the as-prepared polymers were estimated by gel permeation chromatography (GPC, Tosoh Corporation, HPLC–8320 GPC, Jeonju, Korea) equipped with an RI detector. The Fourier transform infrared (FT–IR) analysis was performed to confirm the introduction of functional groups through bromination, chloromethylation, blending, and quaternization reactions. The morphology of the hydrated membranes (Br^-, Cl^- forms) were characterized with the tapping mode using an atomic force microscope (AFM: nanoscope V multimode 8 AFM), installed at the Jeonju Center, Korea Basic Science Research Institute (KBSI), Jeonju, Korea. The thermal behavior of the blended membranes was evaluated using a thermal gravimetric analyzer (TGA, Q 50).

The hydroxide conductivity (σ) was measured using a four-electrode conductivity test bench (SciTech Korea, Jeonju, Korea) linked with a PGZ 301 dynamic EIS voltammeter. The ionic conductivity was calculated from the following formula:

$$\sigma = L/RA \quad (1)$$

where L (cm) denotes the distance between the electrode, R (Ω) denotes the resistance of the membrane, and A (cm^2) denotes the area of the membrane.

The activation energy (E_a) of a membrane was calculated using the following equation:

$$ln\sigma = ln\sigma_0 - (E_a/R \times T) \quad (2)$$

where R is the gas constant and T denotes the Kelvin temperature.

The water uptake of membranes was evaluated as follows. The sample was completely dried in an oven at 80 °C for 48 h, and the weight of the dried membrane (W_{dry}) was measured. Afterward, the membrane was immersed in DI water at 30, 50, 70, and 90 °C for 24 h, and then the weight of the hydrated membrane (W_{wet}) was evaluated immediately. The water uptake of the membrane was calculated using the following equation:

$$\text{Water uptake (\%)} = ((W_{wet} - W_{dry})/W_{dry}) \times 100\% \quad (3)$$

The swelling ratio of the membranes was determined by measuring the in-plane (Δi) and through-plane (Δt) direction before and after immersing the membrane into DI water at 30, 50, 70, and 90 °C for 24 h, respectively. The swelling ratio of the membrane was calculated by the following equation:

$$\text{Swelling ratio (\%)} = ((L_{wet} - L_{dry})/L_{dry}) \times 100\% \quad (4)$$

where L_{wet} and L_{dry} were the lengths (in-plane and through-plane directions) of fully hydrated and dried membranes, respectively.

The ion exchange capacity (IEC) of the Q-PAES/PPO-XY blended membranes were determined using the reverse titration method. The OH^- form membrane (10 × 50 mm) was completely dried under in a vacuum oven, and the weight of the dried membrane (M_{dry}) was measured. Afterward, the membrane was immersed into 50 mL of a 0.05 M HCl solution at room temperature for 48 h. Therefore, the OH^- form was sufficiently replaced by the Cl^- form, then the membrane was removed from the HCl solution. Subsequently, the phenolphthalein indicator was added 3~4 drops to the HCl solution excluding the membrane, and was titrated with a 0.01 M NaOH solution until the solution color changed [44]. The IEC values of the Q-PAES/PPO-XY membranes were calculated using the following equation:

$$\text{IEC (mmol g}^{-1}) = (C_{HCl} \times V_{HCl} - C_{NaOH} \times V_{NaOH})/M_{dry} \quad (5)$$

where C_{HCl} and C_{NaOH} denote the molar concentrations of HCl and NaOH solutions, respectively, and V_{HCl} and V_{NaOH} refer to the volumes of HCl and NaOH solution reaching the equivalent point. The alkaline stability of Q-PAES/PPO-XY blended membranes was monitored by soaking them in a 2.0 M KOH solution at 50 °C for 1000 h. The blended membranes were removed from the alkaline solution, and washed repeatedly with DI water [6]. The FT-IR spectra, IEC, and TGA of the samples were observed at 200 h intervals throughout the alkaline stability test.

4. Discussion

4.1. Characterization of Br-HB-PAES and CM-PPO Polymers

The Br-HB-PAES block copolymer containing a hyperbranched structure was synthesized through three synthetic steps, as shown in Scheme 1. The HB-PAES block copolymer was prepared via direct copolymerization with HB-PAES-Cl and PAES-OH. To introduce the hyperbranched structure in the polymer main chain, the hyperbranched hydrophobic HB-PAES-Cl components were synthesized by copolymerization at a molar ratio of THPE/BPS (1.0:2.2) in the presence of K_2CO_3 as a basic catalyst in an anhydrous system. To optimize the conditions to form the hyperbranched structure of HB-PAES, the polymerization of HB-PAES was conducted under high temperature for a short reaction time to avoid the formation of an undesirable complex network of linear chains, as referred to in a previous report [43].

Subsequently, the HB-PAES block copolymers were brominated using NBS as a bromination agent with BPO as a catalyst. The chemical structure of the as-prepared polymers was characterized by ^1H NMR. As shown in Figure 1, the peaks corresponding to the main backbone were detected at 7.95–6.70 ppm and 2.15–1.46 ppm. After bromination, new proton peaks appeared at 4.47 ppm, which correspond to bromide methyl groups (–CH_2Br). These results indicate that the brominated HB-PAES was successfully synthesized. The degree of bromination (DB) was calculated from the relative integration ratio of the bromo methyl proton groups and the unreacted benzyl methyl proton groups in the main backbone (DB: 42%) [44,45].

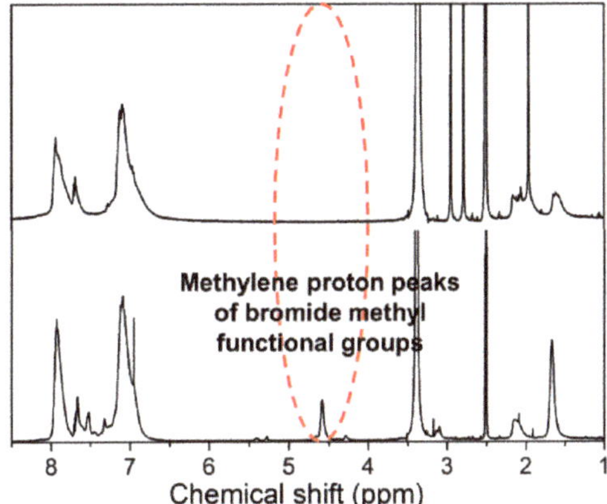

Figure 1. ^1H NMR spectra of the HB-PAES and Br-HB-PAES (The circle indicate the new proton peak in Br-HB-PAES).

The chloromethylated PPO (CM-PPO) was prepared via the Friedel-Crafts alkylation according to a previously reported paper [8,46,47]. As shown in Scheme S1, the chloromethylation reaction of the PPO was carefully conducted to avoid the unexpected side-reactions (gelation, crosslinking, etc.) [46]. To confirm the structure and degree of chloromethylation (DC) of the as-prepared CM-PPO, ^1H NMR was performed (Figure 2). New proton peaks appeared around 4.99–4.67 ppm, which correspond to the chloride methyl groups (–CH_2Cl). The DC was calculated from the relative integration ratio of the chloride methyl proton peaks and unreacted benzyl proton peaks in the main backbone (DC: 51%).

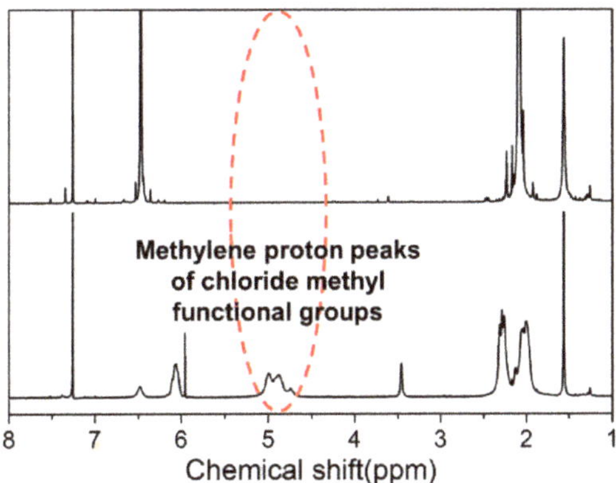

Figure 2. ^1H NMR spectra of the poly(phenylene oxide) (PPO) and chloromethylated poly(phenylene oxide) (CM-PPO) (The circle indicate the new proton peak in CM-PPO).

4.2. Fabrication and Charaterization of Blended Membranes

As listed in Table 1, a series of BC-PAES/PPO-XY blended membranes were prepared by controlling the mixing weight ratio of Br-HB-PAES to Cl-PPO. The fabricated membranes were macroscopically homogeneous and were uniformly mixed to form transparent membranes. The chemical structures of

Br-HB-PAES, CM-PPO, BC-PAES/PPO-55, and Q-PAES/PPO-55 were confirmed by FT-IR, as illustrated in Figure 3. The IR bands of the BC-PAES/PPO-55 blended membrane showed two major peaks at 605 and 1190 cm^{-1}, which are assigned to C-Br and C-Cl stretching vibrations, respectively. In addition, a vibration corresponding to sulfone (S=O) appeared at 1243 cm^{-1}. The absorption peak at 1585 cm^{-1} is ascribed to the stretching vibration of C=C in aromatic benzene rings, indicating that the BC-PAES/PPO-55 blended membrane was successfully blended. The IR spectrum of the Q-PAES/PPO-55 blended membrane includes a peak at 3393 cm^{-1} from the stretching vibration of O-H groups and a peak around 1048 cm^{-1} from the stretching vibration of C-N bands [48,49]. The results show that the blended membranes were successfully fabricated and the quaternization reaction was carried out well.

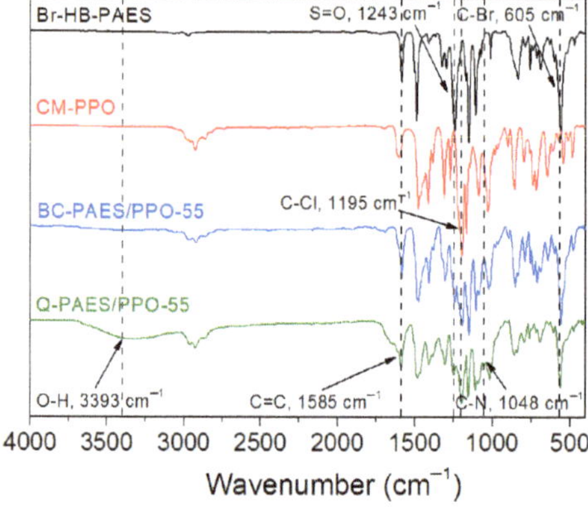

Figure 3. FT-IR spectra of Br-HB-PAES, CM-PPO, BC-PAES/PPO-55 blended membrane, and Q-PAES/PPO-55 blended membrane.

The solubility of the obtained BC-PAES/PPO-XY blended membranes was measured in various organic solvents at 40 °C. As listed in Table 2, the BC-PAES/PPO-XY blended membranes dissolved easily in polar aprotic solvents such as dimethyl sulfoxide (DMSO), *N,N*-dimethylformamide (DMF), tetrahydrofuran (THF), trichloroethylene (TCE), N-methyl-2-pyrrolidone (NMP), and dichloromethane (MC).

Table 2. The solubility of BC-PAES/PPO-XY blending membranes.

Membrane	DMSO	DMF	THF	TCE	NMP	MC	Chloroform	Acetone	Methanol
BC-PAES/PPO-37	++	++	++	++	++	++	++	-	–
BC-PAES/PPO-46	++	++	++	++	++	++	++	-	–
BC-PAES/PPO-55	++	++	++	++	++	++	++	-	–
BC-PAES/PPO-64	++	+	++	++	++	++	+	–	–
BC-PAES/PPO-73	++	++	++	++	++	++	+	–	–

++: Highly soluble; +: Soluble; -: Partially soluble; –: Insoluble.

4.3. Morphology of Q-PAES/PPO-XY Blended Membranes

The micro phase-separation in morphology helps improve the dimensional stability and anion conductivity of AEMs. Therefore, the morphology of AEMs with different blending ratios of Q-HB-PAES and Q-PPO was investigated by the AFM analysis and the results are illustrated in Figure 4. As shown, the dark regions corresponding to hydrophilic areas and the bright regions corresponding to hydrophobic areas were observed, demonstrating that the morphology of all AEMs have the distinct micro phase-separation [10]. It proved that the Q-PAES/PPO-XY blended membranes formed a well-interconnected ion cluster with increasing the Q-PPO block ratio in the blending membrane due to the aggregation of hydrophilic regions resulting from the electrostatic interaction

(N^+/OH^-) and H-bonding interaction (OH^-/H_2O) between Q-HB-PAES and Q-PPO [50–52]. Therefore, the introduction of the hyperbranching network structure in the polymer matrix would promote the enhanced ion network between the ionic groups, which provides a well-connected ion transportation.

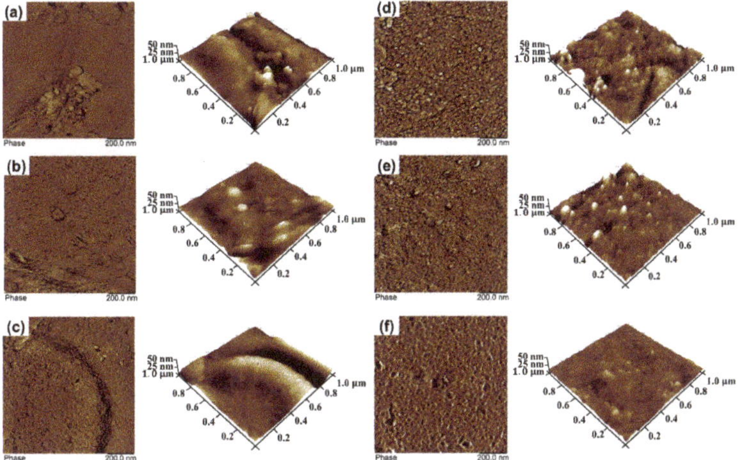

Figure 4. Atomic force microscope (AFM) phase images and three-dimensional (3D) images of (**a**) the pristine Q-HB-PAES membrane, (**b**) Q-PAES/PPO-73, (**c**) Q-PAES/PPO-64, (**d**) Q-PAES/PPO-55, (**e**) Q-PAES/PPO-46, and (**f**) Q-PAES/PPO-37 blended membranes.

4.4. Ionic Exchange Capacity, Water Uptake, and Swelling Ratio

The water uptake and swelling ratio are affected by the IEC values. In general, an increase in the water uptake leads to the increasing capacity to transport OH^- ions in the membrane, improving the IEC values [6,36]. However, the excessive water uptake by the membrane attenuates a dimensional stability [53,54]. The IEC, water uptake, and swelling ratio of the as-prepared membranes were measured at 30, 50, 70, and 90 °C, as presented in Figure 5 and Table 3. The IEC values of the pristine Q-HB-PAES membrane and the Q-PAES/PPO-XY blended membranes were 1.27 and 1.32–1.96 mmol g^{-1}, respectively. In addition, the water uptake and swelling ratio of the Q-PAES/PPO-XY blended membranes exhibited 34.6–50.3% and 20.9–25.0% at 90 °C, respectively. The Q-PAES/PPO-XY blended membranes showed a suitable water uptake, low swelling ratio, and IEC value due to the bulky hyperbranched structure and the relatively strong electrostatic interactions and H-bonding between the blending materials. These membranes can hold more water within the polymer matrix [39,51,52,55,56], compared to the Q-HB-PAES membrane. Furthermore, the IEC value of the Q-PAES/PPO-55 blended membranes, with the most miscible blending weight ratio, was 1.84 mmol g^{-1}.

Table 3. Ionic exchange capacity, water uptake, and swelling ratio of the pristine Q-HB-PAES membrane and the Q-PAES/PPO-XY blended membranes.

Membranes	Water Uptake (%)				Swelling Ratio (%)				IEC (mmol g^{-1})
					Δi		Δt		
	30 °C	50 °C	70 °C	90 °C	30 °C	90 °C	30 °C	90 °C	
Q-HB-PAES	15.5	22.7	32.0	33.2	6.5	17.9	7.1	18.2	1.24 ± 0.03
Q-PAES/PPO-37	24.3	37.1	46.1	50.3	10.2	25.0	12.1	27.1	1.97 ± 0.04
Q-PAES/PPO-46	23.0	35.5	45.0	49.2	9.1	24.9	11.0	28.0	1.90 ± 0.03
Q-PAES/PPO-55	21.3	32.9	41.4	45.4	9.0	22.7	10.7	25.7	1.79 ± 0.05
Q-PAES/PPO-64	21.2	31.3	39.1	42.3	8.7	21.5	9.2	22.9	1.32 ± 0.04
Q-PAES/PPO-73	16.7	23.5	32.5	34.7	7.1	20.9	8.6	21.4	1.27 ± 0.05

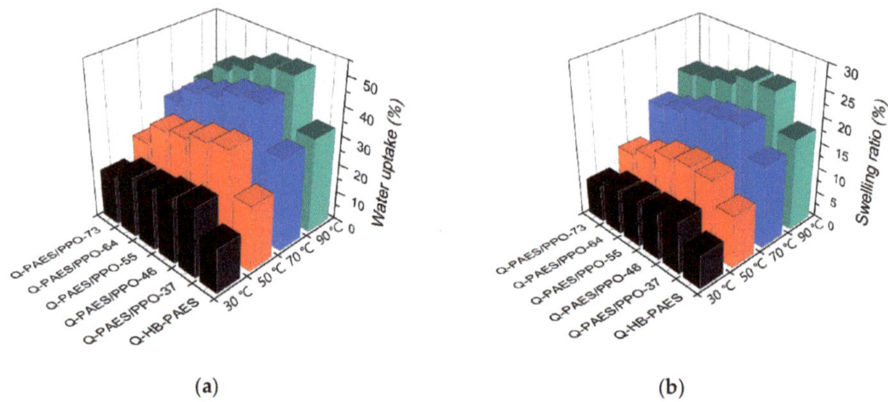

Figure 5. (**a**) Water uptake and (**b**) swelling ratio of the pristine Q-HB-PAES membrane and the Q-PAES/PPO-XY blended membranes.

4.5. Hydroxide Conductivity and Arrhenius Plots of the Q-PAES/PPO-XY Blended Membranes

As listed in Table 4, the hydroxide conductivity of the AEMs, an important parameter, that plays a significant role in fuel cell performance, steadily increased with the temperature due to the enhanced water mobility [28], and was influenced by IEC. High IEC values facilitate the ion transport, which improves the water uptake [29]. The ionic conductivity of the membrane must exceed 10 mS cm^{-1} at RT for the AEM to be applied to fuel cells [28,38]. Figure 6a shows the four-probe hydroxide conductivities of the pristine Q-HB-PAES membrane, and the Q-PAES/PPO-XY blended membranes. The results indicate an increased miscibility between Q-HB-PAES and Q-PPO mediated by the electrostatic interaction and H-bonding interaction, as well as an increased hydroxide conductivity from 25.1–53.9 mS cm^{-1} at 30 °C and from 55.3–90.9 mS cm^{-1} at 90 °C. In particular, the Q-PAES/PPO-55 blended membranes exhibited the highest ionic conductivity value. Due to fact that the balance of two polymers in the blended membranes via electrostatic interactions and H-bonding interactions are increased, it is believed that the Q-PAES/PPO-55 membrane has a more well-defined hydrophilic/hydrophobic microphase separation to form wide ion conducting channels than the other Q-PAES/PPO-XY membranes (37, 46, 64, and 73) [35,51]. As shown in the Arrhenius plots (Figure 6b), the achieved activation energy (E_a) of the as-prepared pristine membrane and blending membranes ranged from 8.03 to 12.80 kJ mol g^{-1}, indicating that the ion transportation mechanism followed the vehicle mechanism [30]. Moreover, the hydroxide conductivity comparison graph with similar IEC values of the recently reported AEMs is shown in Figure 7 [7,8,12,14,19,30,46,55–57].

Table 4. The hydroxide conductivity of the Q-PAES/PPO-XY blended membranes according to the blending weight ratio by temperature.

Membranes	Hydroxide Conductivity (mS cm^{-1})			
	30 °C	50 °C	70 °C	90 °C
Q-HB-PAES	5.1	15.3	26.4	32.8
Q-PAES/PPO-37	28.4	41.5	53.7	62.9
Q-PAES/PPO-46	37.9	53.7	70.5	80.2
Q-PAES/PPO-55	53.9	67.9	80.9	90.9
Q-PAES/PPO-64	32.9	48.9	65.5	75.2
Q-PAES/PPO-73	25.1	36.4	47.4	55.3

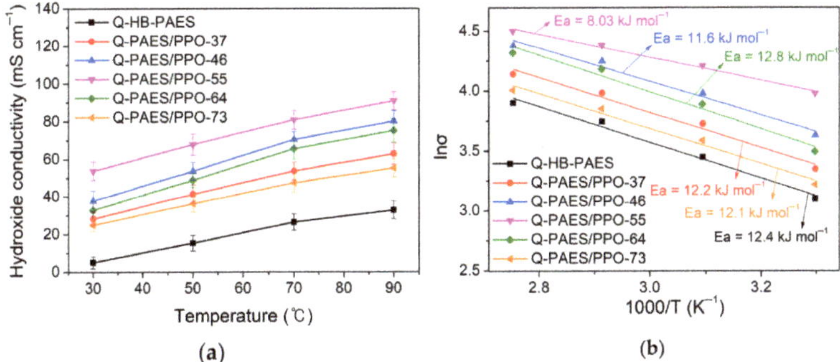

Figure 6. Temperature dependence (30, 50, 70, and 90 °C, 100% RH) of (**a**) hydroxide conductivity and (**b**) Arrhenius plots of the pristine Q-HB-PAES membrane and the Q-PAES/PPO-XY blended membranes.

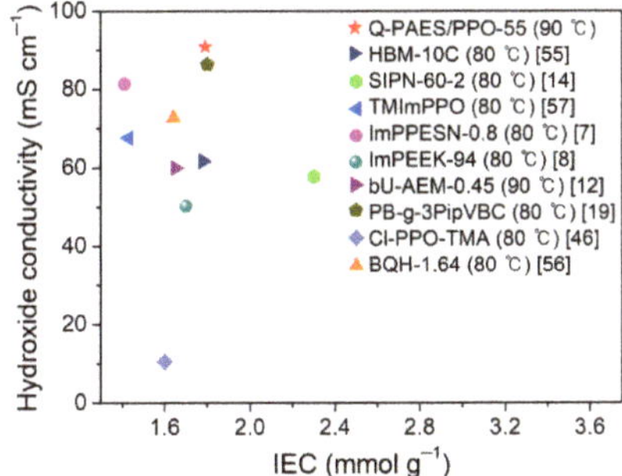

Figure 7. Comparison of hydroxide conductivity and ion exchange capacity (IEC) values of the recently reported anion exchange membranes (AEMs).

4.6. Thermal Properties

The thermal properties of AEMs are important for the practical application of AEMFCs, which are operated between 80–100 °C. As shown in Figure 8, the thermal degradation of the Q-PAES/PPO-XY blended membranes was investigated by TGA at a temperature rising rate of 10 °C min^{-1} in nitrogen atmosphere, which showed three weight loss stages. The initial stage around 135–180 °C is attributed to the volatilization of water absorbed in the membranes. The second stage of weight loss in the range of 180–400 °C corresponds to the decomposition of quaternary ammonium groups. The third stage of weight loss (above 400 °C) is assigned to the decomposition of the polymer main chains [42,58].

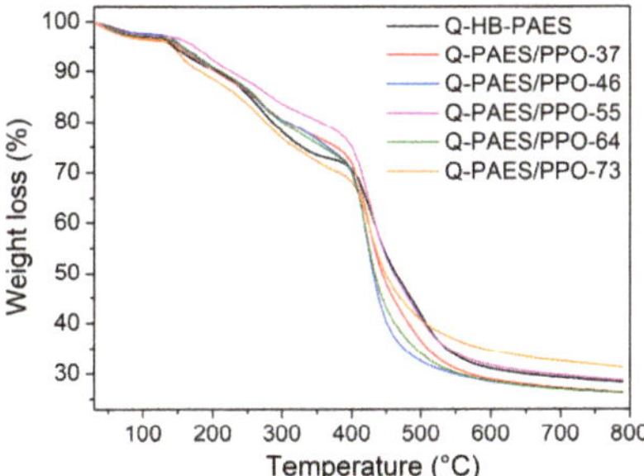

Figure 8. TGA curves of the Q-PAES/PPO-XY blended membranes (temperature range: 30–800 °C, heating ramp rate: 10 °C min^{-1}).

4.7. Alkaline Stability

The alkaline stability of AEMs is a critical parameter, AEMs must operate in high pH environments in FC applications [59–62]. In general, the quaternary ammonium groups of the AEMs disintegrate in a harsh alkaline condition as the result of the OH^- attack via Hofmann elimination and direct nucleophilic substitution [47,63]. In the present work, the Q-PAES/PPO-55 blended membranes achieved the highest electrochemical performance and dimensional stability. Its chemical stability was tested by immersion in a 2.0 M KOH aqueous solution at 50 °C for 1000 h, followed by the evaluation of its FT-IR spectra and IEC. As shown in Figure 9, the transition of the chemical structure of the AEMs resulting from the alkaline stability test was characterized via the FT-IR spectra. A small change around 3393 cm^{-1} was observed, corresponding to the O-H stretching vibration, and there was a slight change of the peak at 1048 cm^{-1} corresponding to the C-N stretching vibration. However, there was no significant visible change in the overall spectrum. Moreover, at 200 h intervals during the prolonged alkaline treatment, the Q-PAES/PPO-55 blended membranes were tested by the IEC titration method. It turns out that the IEC values remain at 85.3% during immersion in a high pH environment for the alkaline stability test, as shown in Figure 10. The assembly of the bulky hyperbranched structure and the linear structure protects the backbone and quaternary ammonium groups from the OH^- attack through the steric hindrance effect. The result showed that chemically stable Q-PAES/PPO-XY blended membranes were obtained [31,33,55].

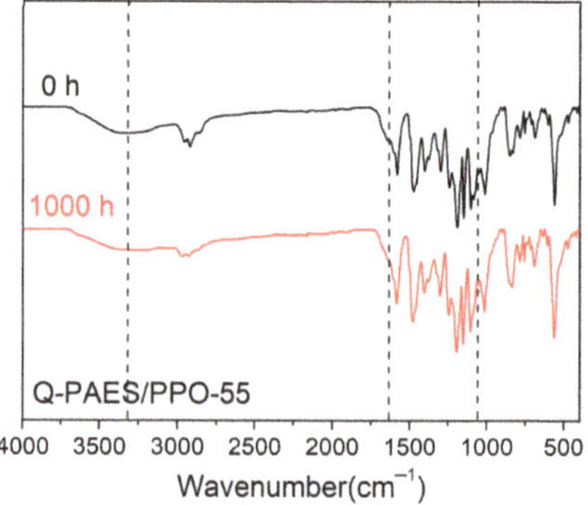

Figure 9. FT-IR spectra of Q-PAES/PPO-55 blended membranes before and after the alkaline stability test (immersed in a 2.0 M KOH aqueous solution at 50 °C for 1000 h).

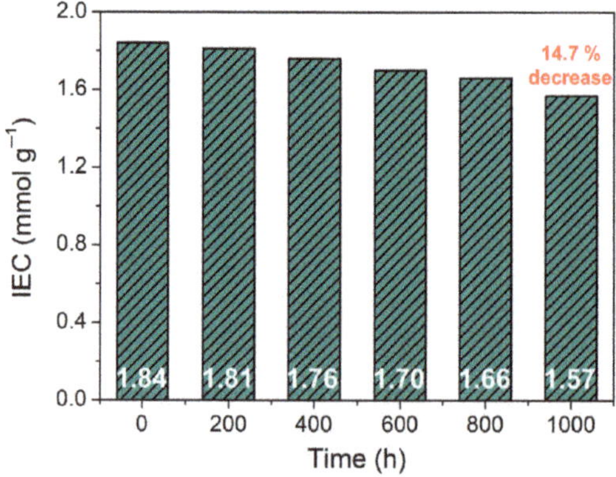

Figure 10. IEC of Q-PAES/PPO-55 blended membranes before and after the alkaline stability test (immersed in a 2.0 M KOH aqueous solution at 50 °C for 1000 h).

5. Conclusions

In summary, a series of Q-PAES/PPO-XY blended membranes were prepared by simply blending hyperbranched brominated PAES and linear chloromethylated PPO different weight ratios (30, 40, 50, 60, and 70 wt%) to improve the hydroxide conductivity and dimensional stability of AEMs. The Q-PAES/PPO-XY membranes form electrostatic interaction and hydrogen bonding networks between the quaternized functional groups introduced in the HB-PAES and PPO, and the blended membranes showed excellent dimensional/chemical properties, depending on the mixing ratio of the polymers, compared to the pristine Q-HB-PAES membrane. The dense and concentrated ammonium groups promoted the nanophase separation of the AEM and showed excellent electrochemical properties despite the low IEC. Among the blended membranes, the Q-PAES/PPO-55 membrane achieved the highest hydroxide ion conductivity (90.9 mS cm^{-1} at 90 °C) and showed a reasonable alkaline stability under high pH conditions. Therefore, the developed blended membranes are a promising candidate for AEM fuel cell applications.

Supplementary Materials: The following are available online at http://www.mdpi.com/2073-4360/12/12/3011/s1, Figure S1: ^{1}H NMR spectra (a) hydrophobic oligomer (HB-PAES-Cl), (b) hydrophilic precursor (PAES-OH), and (c) block copolymer (HB-PAES), Scheme S1: Synthesis process of chloromethylated poly(phenylene oxide) (CM-PPO), Figure S2: In-plane swelling ratios and through-plane swelling ratios of (1) Q-HB-PAES pristine membrane, (2) Q-PAES/PPO-37, (3) Q-PAES/PPO-46, (4) Q-PAES/PPO-55, (5) Q-PAES/PPO-64, and (6) Q-PAES/PPO-73 blended membranes.

Author Contributions: Investigation, visualization, writing—original draft preparation, S.H.K.; writing—original draft preparation, writing—review and editing, supervision, validation, K.H.L. and J.Y.C.; supervision, data curation, validation, A.R.K.; resources, project administration, funding acquisition, supervision, D.J.Y. All authors have read and agreed to the published version of the manuscript.

Funding: This work was supported by the research funds of the Jeonbuk National University, Republic of Korea in 2020. This work was supported by the Korea Institute of Energy Technology Evaluation and Planning (KETEP) and the Ministry of Trade, Industry & Energy (MOTIE) of the Republic of Korea (no. 20184030202210). This work was supported by grants from the Medical Research Center Program (NRF-2017R1A5A2015061) through the National Research Foundation (NRF), which is funded by the Korean government (MSIP).

Conflicts of Interest: The authors declare no conflict of interest.

References

1. Vijayakumar, V.; Nam, S.Y. Recent advancements in applications of alkaline anion exchange membranes for polymer electrolyte fuel cells. *J. Ind. Eng. Chem.* **2019**, *70*, 70–86. [CrossRef]
2. Li, N.; Guiver, M.D. Ion transport by nanochannels in ion-containing aromatic copolymers. *Macromolecules* **2014**, *47*, 2175–2198. [CrossRef]
3. Ramanavicius, S.; Ramanavicius, A. Progress and insights in the application of MXenes as new 2D nano-materials suitable for biosensors and biofuel cell design. *Int. J. Mol. Sci.* **2020**, *21*, 9224. [CrossRef] [PubMed]
4. Lee, K.H.; Chu, J.Y.; Kim, A.R.; Yoo, D.J. Effect of functionalized SiO_2 toward proton conductivity of composite membranes for PEMFC application. *Int. J. Energy Res.* **2019**, *43*, 5333–5345. [CrossRef]
5. Yoo, D.J.; Hyun, S.H.; Kim, A.R.; Kumar, G.G.; Nahm, K.S. Novel sulfonated poly(arylene biphenylsulfone ether) copolymers containing bisphenylsulfonyl biphenyl moiety: Structural, thermal, electrochemical and morphological characteristics. *Polym. Int.* **2011**, *60*, 85–92. [CrossRef]
6. Chu, J.Y.; Lee, K.H.; Kim, A.R.; Yoo, D.J. Improved electrochemical performance of composite anion exchange membranes for fuel cells through cross linking of the polymer chain with functionalized graphene oxide. *J. Membr. Sci.* **2020**, *611*, 118385. [CrossRef]
7. Lai, A.N.; Zhuo, Y.Z.; Lin, C.X.; Zhang, Q.G.; Zhu, A.M.; Ye, M.L.; Liu, Q.L. Side-chain-type phenolphthalein-based poly(arylene ether sulfone nitrile)s anion exchange membrane for fuel cells. *J. Membr. Sci.* **2016**, *502*, 94–105. [CrossRef]
8. Li, Z.; Jiang, Z.; Tian, H.; Wang, S.; Zhang, B.; Cao, Y.; He, G.; Li, Z.; Wu, H. Preparing alkaline anion exchange membrane with enhanced hydroxide conductivity via blending imidazolium-functionalized and sulfonated poly(ether ether ketone). *J. Power Sources* **2015**, *288*, 384–392. [CrossRef]

9. Sun, Z.; Lin, B.; Yan, F. Anion-exchange membranes for alkaline fuel-cell applications: The effects of cations. *ChemSusChem* **2018**, *11*, 58–70. [CrossRef] [PubMed]
10. Zhang, S.; Zhu, X.; Jin, C. Development of a high-performance anion exchange membrane using poly(isatin biphenylene) with flexible heterocyclic quaternary ammonium cations for alkaline fuel cells. *J. Mater. Chem. A* **2019**, *7*, 6883–6893. [CrossRef]
11. He, S.; Liu, L.; Wang, X.; Zhang, S.; Guiver, M.D.; Li, N. Azide-assisted self-crosslinking of highly ion conductive anion exchange membranes. *J. Membr. Sci.* **2016**, *509*, 48–56. [CrossRef]
12. Wei, H.; Tong, L.; Yu, S.; Zhang, J.; Dong, Y.; Li, X.; Ding, Y. Non-covalently crosslinked anion exchange membranes: Effect of urea hydrogen-bonding group position. *Polymer* **2019**, *179*, 121654. [CrossRef]
13. Qaisrani, N.A.; Ma, L.; Hussain, M.; Liu, J.; Li, L.; Zhou, R.; Jia, Y.; Zhang, F.; He, G. Hydrophilic flexible ether containing, cross-linked anion-exchange membrane quaternized with DABCO. *ACS Appl. Mater. Interfaces* **2020**, *12*, 3510–3521. [CrossRef]
14. Pan, J.; Zhu, L.; Han, J.; Hickner, M.A. Mechanically tough and chemically stable anion exchange membranes from rigid-flexible semi-interpenetrating networks. *Chem. Mater.* **2015**, *27*, 6689–6698. [CrossRef]
15. Kim, A.R.; Vinothkannan, M.; Yoo, D.J. Fabrication of binary sulfonated poly ether sulfone and sulfonated polyvinylidene fluoride-co-hexafluoro propylene blend membrane as efficient electrolyte for proton exchange membrane fuel cells. *Bull. Korean Chem. Soc.* **2018**, *39*, 913–919. [CrossRef]
16. Msomi, P.; Nonjola, P.; Ndungu, P.; Ramontja, J. Quaternized poly(2,6 dimethyl–1,4 phenylene oxide)/polysulfone blended anion exchange membrane for alkaline fuel cells application. *Mater. Today Proc.* **2018**, *5*, 10496–10504. [CrossRef]
17. Morandi, C.G.; Peach, R.; Krieg, H.M.; Kerres, J. Novel morpholinium-functionalized anion-exchange PBI–polymer blends. *J. Mater. Chem. A* **2015**, *3*, 1110–1120. [CrossRef]
18. Msomi, P.F.; Nonjola, P.T.; Ndungu, P.G.; Ramontja, J. Poly(2, 6-dimethyl-1, 4-phenylene)/polysulfone anion exchange membrane blended with TiO_2 with improved water uptake for alkaline fuel cell application. *Int. J. Hydrogen Energy* **2020**, *45*, 29465–29476. [CrossRef]
19. Kim, A.R.; Vinothkannan, M.; Song, M.H.; Lee, J.-Y.; Lee, H.-K.; Yoo, D.J. Amine functionalized carbon nanotube (ACNT) filled in sulfonated poly(ether ether ketone) membrane: Effects of ACNT in improving polymer electrolyte fuel cell performance under reduced relative humidity. *Compos. Part B Eng.* **2020**, *188*, 107890. [CrossRef]
20. Nhung, L.T.; Kim, I.Y.; Yoon, Y.S. Quaternized chitosan-based anion exchange membrane composited with quaternized poly(vinylbenzyl chloride)/polysulfone blend. *Polymers* **2020**, *12*, 2714. [CrossRef]
21. Li, N.; Yan, T.; Li, Z.; Thurn-Albrecht, T.; Binder, W.H. Comb-shaped polymers to enhance hydroxide transport in anion exchange membranes. *Energy Environ. Sci.* **2012**, *5*, 7888–7892. [CrossRef]
22. Li, N.; Wang, L.; Hickner, M. Cross-linked comb-shaped anion exchange membranes with high base stability. *Chem. Commun.* **2014**, *50*, 4092–4095. [CrossRef] [PubMed]
23. Zhang, S.; Wang, Y.; Gao, X.; Liu, P.; Wang, X.; Zhu, X. Enhanced conductivity and stability via comb-shaped polymer anion exchange membrane incorporated with porous polymeric nanospheres. *J. Membr. Sci.* **2020**, *597*, 117750. [CrossRef]
24. Matsumura, S.; Hlil, A.R.; Du, N.; Lepiller, C.; Gaudet, J.; Guay, D.; Shi, Z.; Holdcroft, S.; Hay, A.S. Ionomers for proton exchange membrane fuel cells with sulfonic acid groups on the end-groups: Novel branched poly(ether-ketone)s with 3,6-ditrityl-9H-carbazole end-groups. *J. Polym. Sci. Part A Polym. Chem.* **2008**, *46*, 3860–3868. [CrossRef]
25. Gao, X.L.; Yang, Q.; Wu, H.Y.; Sun, Q.H.; Zhu, Z.Y.; Zhang, Q.G.; Zhu, A.M.; Liu, Q.L. Orderly branched anion exchange membranes bearing long flexible multi-cation side chain for alkaline fuel cells. *J. Membr. Sci.* **2019**, *589*, 117247. [CrossRef]
26. Wang, K.; Wu, Q.; Yan, X.; Liu, J.; Gao, L.; Hu, L.; Zhang, N.; Pan, Y.; Zheng, W.; He, G. Branched poly(ether ether ketone) based anion exchange membrane for H_2/O_2 fuel cell. *Int. J. Hydrogen Energy* **2019**, *44*, 23750–23761. [CrossRef]
27. Pan, J.; Li, Y.; Zhuang, L.; Lu, J. Self-crosslinked alkaline polymer electrolyte exceptionally stable at 90 °C. *Chem. Commun.* **2010**, *46*, 8597–8599. [CrossRef]
28. Robertson, N.J.; Kostalik, H.A.; Clark, T.J.; Mutolo, P.F.; Abruña, H.D.; Coates, G.W. Tunable high performance cross-linked alkaline anion exchange membranes for fuel cell applications. *J. Am. Chem. Soc.* **2010**, *132*, 3400–3404. [CrossRef]

29. Ge, Q.; Liu, Y.; Yang, Z.; Wu, B.; Hu, M.; Liu, X.; Hou, J.; Xu, T. Hyper-branched anion exchange membranes with high conductivity and chemical stability. *Chem. Commun.* **2016**, *52*, 10141–10143. [CrossRef]
30. Lee, K.H.; Cho, D.H.; Kim, Y.M.; Moon, S.J.; Seong, J.G.; Shin, D.W.; Sohn, J.-Y.; Kim, J.F.; Lee, Y.M. Highly conductive and durable poly(arylene ether sulfone) anion exchange membrane with end-group cross-linking. *Energy Environ. Sci.* **2017**, *10*, 275–285. [CrossRef]
31. Ghanem, A.F.; El-Gendi, A.; Rehim, M.H.A.; El-Khatib, K.M. Hyperbranched polyester and its sodium titanate nanocomposites as proton exchange membranes for fuel cells. *RSC Adv.* **2016**, *6*, 32245–32257. [CrossRef]
32. Chu, P.P.; Wu, C.-S.; Liu, P.-C.; Wang, T.-H.; Pan, J.-P. Proton exchange membrane bearing entangled structure: Sulfonated poly(ether ether ketone)/bismaleimide hyperbranch. *Polymer* **2010**, *51*, 1386–1394. [CrossRef]
33. Georgi, U.; Erber, M.; Stadermann, J.; Abulikemu, M.; Komber, H.; Lederer, A.; Voit, B. New approaches to hyperbranched poly(4-chloromethylstyrene) and introduction of various functional end groups by polymer-analogous reactions. *J. Polym. Sci. Part A Polym. Chem.* **2010**, *48*, 2224–2235. [CrossRef]
34. Gong, S.; Li, L.; Ma, L.; Qaisrani, N.A.; Liu, J.; He, G.; Zhang, F. Blend anion exchange membranes containing polymer of intrinsic microporosity for fuel cell application. *J. Membr. Sci.* **2020**, *595*, 117541. [CrossRef]
35. Kim, A.R.; Vinothkannan, M.; Yoo, D.J. Sulfonated-fluorinated copolymer blending membranes containing SPEEK for use as the electrolyte in polymer electrolyte fuel cells (PEFC). *Int. J. Hydrogen Energy* **2017**, *42*, 4349–4365. [CrossRef]
36. Kim, A.R.; Park, C.J.; Vinothkannan, M.; Yoo, D.J. Sulfonated poly ether sulfone/heteropoly acid composite membranes as electrolytes for the improved power generation of proton exchange membrane fuel cells. *Compos. Part B Eng.* **2018**, *155*, 272–281. [CrossRef]
37. Ran, J.; Wu, L.; Ru, Y.; Hu, M.; Din, L.; Xu, T. Anion exchange membranes (AEMs) based on poly(2,6-dimethyl-1,4-phenylene oxide) (PPO) and its derivatives. *Polym. Chem.* **2015**, *6*, 5809–5826. [CrossRef]
38. Zhu, Y.; Ding, L.; Liang, X.; Shehzad, M.A.; Wang, L.; Ge, X.; He, Y.; Wu, L.; Varcoe, J.R.; Xu, T. Beneficial use of rotatable-spacer side-chains in alkaline anion exchange membranes for fuel cells. *Energy Environ. Sci.* **2018**, *11*, 3472–3479. [CrossRef]
39. Martín-Zarco, M.; Titvinidze, G.; García-Martínez, J.C.; Rodríguez-López, J. Sulfonated dendrimer- and hyperbranched polyglycerol-PBIOO® blend membranes for fuel cells. *J. Polym. Sci. Part A Polym. Chem.* **2016**, *54*, 69–80. [CrossRef]
40. Gong, S.; Bai, L.; Li, L.; Qaisrani, N.A.; Ma, L.; He, G.; Zhang, F. Block copolymer anion exchange membrane containing polymer of intrinsic microporosity for fuel cell application. *Int. J. Hydrogen Energy* **2020**, in press. [CrossRef]
41. Chu, J.Y.; Lee, K.H.; Kim, A.R.; Yoo, D.J. Study on the chemical stabilities of poly(arylene ether) random copolymers for alkaline fuel cells: Effect of main chain structures with different monomer units. *ACS Sustain. Chem. Eng.* **2019**, *7*, 20077–20087. [CrossRef]
42. Ma, L.; Qaisrani, N.A.; Hussain, M.; Li, L.; Jia, Y.; Ma, S.; Zhou, R.; Bai, L.; He, G.; Zhang, F. Cyclodextrin modified, multication cross-linked high performance anion exchange membranes for fuel cell application. *J. Membr. Sci.* **2020**, *607*, 118190. [CrossRef]
43. Kim, J.H.; Vinothkannan, M.; Kim, A.R.; Yoo, D.J. Anion exchange membranes obtained from poly(arylene ether sulfone) block copolymers comprising hydrophilic and hydrophobic segments. *Polymers* **2020**, *12*, 325. [CrossRef]
44. Zhang, M.; Liu, J.; Wang, Y.; An, L.; Guiver, M.D.; Li, N. Highly stable anion exchange membranes based on quaternized polypropylene. *J. Mater. Chem. A* **2015**, *3*, 12284–12296. [CrossRef]
45. Yan, J.; Zhu, L.; Chaloux, B.L.; Hickner, M.A. Anion exchange membranes by bromination of tetramethylbiphenol-based poly(sulfone)s. *Polym. Chem.* **2017**, *8*, 2442–2449. [CrossRef]
46. Becerra-Arciniegas, R.A.; Narducci, R.; Ercolani, G.; Antonaroli, S.; Sgreccia, E.; Pasquini, L.; Knauth, P.; Di Vona, M.L. Alkaline stability of model anion exchange membranes based on poly(phenylene oxide) (PPO) with grafted quaternary ammonium groups: Influence of the functionalization route. *Polymer* **2019**, *185*, 121931. [CrossRef]
47. Lin, B.; Xu, F.; Chu, F.; Ren, Y.; Ding, J.; Yan, F. Bis-imidazolium based poly(phenylene oxide) anion exchange membranes for fuel cells: The effect of cross-linking. *J. Mater. Chem. A* **2019**, *7*, 13275–13283. [CrossRef]

48. Alam, M.M.; Wang, Y.; Jiang, C.; Xu, T.; Liu, Y.; Xu, T. A novel anion exchange membrane for bisulfite anion separation by grafting a quaternized moiety through BPPO via thermal-induced phase separation. *Int. J. Mol. Sci.* **2020**, *21*, 5782. [CrossRef]
49. Fang, J.; Shen, P.K. Quaternized poly(phthalazinon ether sulfone ketone) membrane for anion exchange membrane fuel cells. *J. Membr. Sci.* **2006**, *285*, 317–322. [CrossRef]
50. McNair, R.; Cseri, L.; Szekely, G.; Dryfe, R. Asymmetric membrane capacitive deionization using anion-exchange membranes based on quaternized polymer blends. *Acs Appl. Polym. Mater.* **2020**, *2*, 2946–2956. [CrossRef] [PubMed]
51. Chen, W.; Hu, M.; Wang, H.; Wu, X.; Gong, X.; Yan, X.; Zhen, D.; He, G. Dimensionally stable hexamethylenetetramine functionalized polysulfone anion exchange membranes. *J. Mater. Chem. A* **2017**, *5*, 15038–15047. [CrossRef]
52. Chen, C.; Tse, Y.-L.S.; Lindberg, G.E.; Knight, C.; Voth, G.A. Hydroxide solvation and transport in anion exchange membranes. *J. Am. Chem. Soc.* **2016**, *138*, 991–1000. [CrossRef]
53. Zhao, Y.; Yoshimura, K.; Shishitani, H.; Yamaguchi, S.; Tanaka, H.; Koizumi, S.; Szekely, N.; Radulescu, A.; Richter, D.; Maekawa, Y. Imidazolium-based anion exchange membranes for alkaline anion fuel cells: Elucidation of the morphology and the interplay between the morphology and properties. *Soft Matter* **2016**, *12*, 1567–1578. [CrossRef]
54. Abouzari-lotf, E.; Ghassemi, H.; Nasef, M.M.; Ahmad, A.; Zakeri, M.; Ting, T.M.; Abbasi, A.; Mehdipour-Ataei, S. Phase separated nanofibrous anion exchange membranes with polycationic side chains. *J. Mater. Chem. A* **2017**, *5*, 15326–15341. [CrossRef]
55. Ge, Q.; Liang, X.; Ding, L.; Hou, J.; Miao, J.; Wu, B.; Yang, Z.; Xu, T. Guiding the self-assembly of hyperbranched anion exchange membranes utilized in alkaline fuel cells. *J. Membr. Sci.* **2019**, *573*, 595–601. [CrossRef]
56. Niu, M.; Zhang, C.; He, G.; Zhang, F.; Wu, X. Pendent piperidinium-functionalized blend anion exchange membrane for fuel cell application. *Int. J. Hydrogen Energy* **2019**, *44*, 15482–15493. [CrossRef]
57. Zhu, Y.; He, Y.; Ge, X.; Liang, X.; Shehzad, M.A.; Hu, M.; Liu, Y.; Wu, L.; Xu, T. A benzyltetramethylimidazolium-based membrane with exceptional alkaline stability in fuel cells: Role of its structure in alkaline stability. *J. Mater. Chem. A* **2018**, *6*, 527–534. [CrossRef]
58. Kang, D.H.; Das, G.; Yoon, H.H.; Kim, I.T. A composite anion conducting membrane based on quaternized cellulose and poly(phenylene oxide) for alkaline fuel cell applications. *Polymers* **2020**, *12*, 2676. [CrossRef] [PubMed]
59. Chen, J.; Li, C.; Wang, J.; Li, L.; Wei, Z. A general strategy to enhance the alkaline stability of anion exchange membranes. *J. Mater. Chem. A* **2017**, *5*, 6318–6327. [CrossRef]
60. Jung, M.-s.J.; Arges, C.G.; Ramani, V. A perfluorinated anion exchange membrane with a 1,4-dimethylpiperazinium cation. *J. Mater. Chem.* **2011**, *21*, 6158–6160. [CrossRef]
61. Arges, C.G.; Parrondo, J.; Johnson, G.; Nadhan, A.; Ramani, V. Assessing the influence of different cation chemistries on ionic conductivity and alkaline stability of anion exchange membranes. *J. Mater. Chem.* **2012**, *22*, 3733–3744. [CrossRef]
62. Son, T.Y.; Kim, T.-H.; Nam, S.Y. Crosslinked pore-filling anion exchange membrane using the cylindrical centrifugal force for anion exchange membrane fuel cell system. *Polymers* **2020**, *12*, 2758. [CrossRef] [PubMed]
63. Qiu, X.; Ueda, M.; Fang, Y.; Chen, S.; Hu, Z.; Zhang, X.; Wang, L. Alkaline stable anion exchange membranes based on poly(phenylene-co-arylene ether ketone) backbones. *Polym. Chem.* **2016**, *7*, 5988–5995. [CrossRef]

Article

Modifying the Catalyst Layer Using Polyvinyl Alcohol for the Performance Improvement of Proton Exchange Membrane Fuel Cells under Low Humidity Operations

Prathak Jienkulsawad [1], Yong-Song Chen [2,*] and Amornchai Arpornwichanop [1]

[1] Center of Excellence in Process and Energy Systems Engineering, Department of Chemical Engineering, Faculty of Engineering, Chulalongkorn University, Bangkok 10330, Thailand; prathak.j@gmail.com (P.J.); amornchai.a@chula.ac.th (A.A.)

[2] Department of Mechanical Engineering and Advanced Institute of Manufacturing with High-tech Innovations, National Chung Cheng University, Chiayi County 62102, Taiwan

* Correspondence: imeysc@ccu.edu.tw

Received: 30 July 2020; Accepted: 16 August 2020; Published: 19 August 2020

Abstract: A proton exchange membrane fuel cell (PEMFC) system for the application of unmanned aerial vehicles is equipped without humidifiers and the cathode channels of the stack are open to the environment due to limited weight available for power sources. As a result, the PEMFC is operated under low humidity conditions, causing membrane dehydration, low performance, and degradation. To keep the generated water within the fuel cell to humidify the membrane, in this study, polyvinyl alcohol (PVA) is employed in the fabrication of membrane electrode assemblies (MEAs). The effect of PVA content, either sprayed on the gas diffusion layer (GDL) or mixed in the catalyst layer (CL), on the MEA performance is compared under various humidity conditions. The results show that MEA performance is increased with the addition of PVA either on the GDL or in the CL, especially for non-humidified anode conditions. The result suggested that 0.03% PVA in the anode CL and 0.1% PVA on the GDL can improve the MEA performance by approximately 30%, under conditions of a non-humidified anode and a room-temperature-humidified cathode. However, MEAs with PVA in the anode CL show better durability than those with PVA on the GDL according to measurement with electrochemical impedance spectroscopy.

Keywords: proton exchange membrane fuel cell; humidifier; membrane electrode assembly; polyvinyl alcohol

---

1. Introduction

Proton exchange membrane fuel cells (PEMFCs) have gained much attention for transportation applications, especially in unmanned aerial vehicles (UAVs) [1], due to their low operating temperature (<100 °C), high power density, and high energy density. Presently, as the commercial PEMFC uses a Nafion membrane which needs to be hydrated for facilitating proton transfer, a humidification for the supplied air is recommended to avoid membrane drying [2]. However, carrying a humidifier is a challenge for small UAVs (e.g., multirotors), which have space and weight limits. In this case, the membrane might have to be hydrated by moisture in the air, which is relatively low at the temperature of the PEMFC, or by the water generated during the electrochemical reaction. Many researchers have attempted to modify the membrane electrode assembly (MEA) of the PEMFC and enhance its performance under a low humidity condition or self-humidification [3].

A way to improve MEAs for low humidity operation involves adding some additives into either the membrane or the catalyst layer (CL). The additives should have the ability to hold water, which could be hydrophilic materials, such as SiO_2 [4], TiO_2 [5] or polyvinyl alcohol (PVA) [6], have hydrophilic functional groups such as triazole [7] and vinyl phosphonic acid (VPA) [8], or be a metal-organic framework (MOF) [9]. In addition to mixing the additive into the CL or membrane composite, use of dual CLs has been investigated [10]. Moreover, functionalizing a carbon electrode can also enhance the hydrophilic property of the electrode [11,12]. However, the process of CL modification seems to be simpler compared to that of membrane modification. SiO_2 is mostly used as an additive in the CL in many research investigations due to the high ability of silica to increase the water uptake. Su et al. [13] successfully added SiO_2 into the CL on a Nafion 212 membrane by using the organic colloid method for catalyst ink preparation and the illumination method for spraying and forming a CL. The results of the analysis showed excellent bonding among silica, Pt, and C. No significant difference in the performance of the PEMFC using the anode with or without SiO_2 under external humidification condition was observed, indicating that SiO_2 did not affect the electrochemical reaction. However, a long-term operation (20 h) of PEMFCs with 3 wt.% SiO_2 in the anode showed a drop in performance of 22% with a very stable power of 438 mW cm^{-2}. Han et al. [14] fabricated MEAs with 6 wt.% SiO_2 by using the hot press method to make contact between CLs and the membrane. It was found that the performance of the PEMFC was lower than that reported by Su et al. [13], even though the higher SiO_2 content (6 wt.%) was used. Hence, the method of MEA fabrication has a significant impact on cell performance. MEAs with the catalyst coated membrane (CCM) show better performance than those with the catalyst coated GDL. These results were validated by the study of Leimin et al. [15]. The CCM under irradiation (CSMUI) method showed better contact between the catalyst and membrane, resulting in small cell resistances (total ohmic resistance and charge transfer resistance), compared to the catalyst coated GDL method.

PVA has been widely used in membrane modification because of its high water absorption capability and electrical resistance [6]. However, PVA has low proton conductivity and thus phosphorylated PVA, which has a crosslinking structure, was often used, as it has relatively high proton conductivity and still has water retention ability [16]. El-Toony et al. [17] cast a membrane of phosphorylated polyvinyl alcohol (p-PVA)/poly hydroxybutyrate (PHB) for PEMFCs using the gamma irradiation method for making p-PVA. The membrane was cast in polystyrene petri dishes (solution casting method) and the CLs were assembled with the membrane by the hot-pressed method. The cast membrane showed superior performance compared to Nafion 212 under a relative humidity of 100% in both short- and long-term tests. The maximum power density achieved was about 639 mW cm^{-2}. However, the PEMFC was not tested under a low relative humidity condition in their study. Attaran et al. [18] cast PVA/polyvinyl pyrrolidone (PVP)/$BaZrO_3$ by using the solution casting method. Glutaraldehyde solution, the crosslinking agent, was used to prepare the PVA. The ceramic material $BaZrO_3$, with perovskite structure, was added to further improve proton conductivity. The painting method was applied to coat catalyst on carbon paper, and then the hot-pressing method was applied to assemble the MEA. Although the proton conductivity was improved, the cell performance degraded (28.98 mW cm^{-2}). In addition to adding PVA into a membrane composite, Liang et al. [19] added PVA into the anode CL. They used the illuminated spraying method on the Nafion 212, with a controlled active area of 5 cm^2. The MEA with PVA 5 wt.% was found to be optimal to achieve the maximum power density of 623.3 mW cm^{-2}, with relatively low ohmic resistance and the lowest charge transfer of the MEAs under low humidification (RH 34%) and pressurized condition (20 psi). PVA in CLs shows a remarkable high performance, similar to PVA in the membrane (with the appropriate preparation method), and it is comparative to other additives.

Using PVA as a water absorbent in the anode CL has been investigated at the pressured condition of 20 psi; however, the effect of PVA on the MEA performance under ambient environment needs to be further studied for UAV applications. In this study, performance of MEAs with various PAV concentrations in the anode CL (named the PA method) is evaluated under various cell temperatures and humidifier temperatures. Moreover, coating a thin layer of PVA on the anode GDL (named the PG method) as a water reservoir for hydrating the membrane is proposed in this study. The durability of these MEAs with different configurations is also studied using electrochemical impedance spectroscopy.

2. Experimental

2.1. MEA Fabrication

The MEA consists of a CCM sandwiched between two GDLs (GDL260, CeTech, Taichung City, Taiwan). Catalyst ink was a mixture of 46.8 wt% Pt on carbon (TEC10E50E, Tanaka, Japan), polyvinyl alcohol (72000 BioChemica, AppliChem GmbH, Darmstadt, Germany), Nafion dispersion (D520, Chemours Fluoroproducts), ethanol (95% ethanol, TTL Taiwan), and deionized water. The catalyst ink was mixed in a planetary centrifugal mixer (Thinky mixer, ARE-310, Tokyo, Japan) for 30 min, followed by an ultrasonic bath (Delta D80, Taipei, Taiwan) for 10 min, and again mixed in the ARE-310 for 30 min. The catalyst ink was sprayed on both sides of the membrane (Nafion NR211, Dupont, Wilmington, DE, USA) with an active area of 26.01 cm^2 by using an ultrasonic spraying system (Benchtop BT, USI, USA). The solvent was evaporated during spraying process at 100 °C by using a hot plate (PC-400D, Corning, New York, NY, USA).

The Nafion to carbon ratio in the catalyst ink was predetermined as 1:1 though a preliminary study to study the effect of the PVA location on the performance of the PEMFC under low humidity operation. Pt loadings in the anode and cathode CLs were controlled to be 0.1 mg cm^{-2} and 0.3 mg cm^{-2}, respectively. PVA (0.1 wt.% solution) was prepared by dissolving PVA monomer in DI water at 90 °C. PVA in the CL was 0.01, 0.03, 0.07 and 0.1 wet weight%, using the PA method, based on a preliminary trial. Cathode catalyst ink was prepared by the same method, except no PVA was added. Regarding the PG method, PVA solution was sprayed on the GDL, while the anode catalyst without PVA was coated on the membrane. The amount of PVA on the GDL was controlled as done for the anode CL (the PA method), to study the effect of the different preparation methods. The MEA was assembled by sandwiching the CCM between two GDLs without a hot-pressing step and a Teflon gasket with a thickness of 0.225 mm was used to prevent gas leakage. The schematic of MEAs of a PEMFC, for this study, is shown in Figure 1.

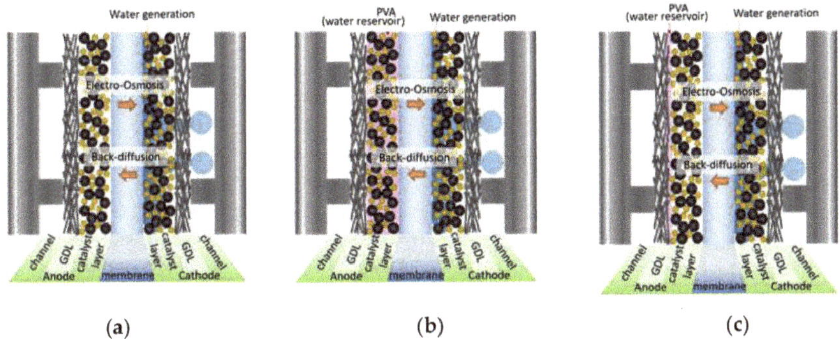

Figure 1. Schematic of a single cell with different electrode configurations: (**a**) no polyvinyl alcohol (PVA) case, (**b**) PVA in the anode catalyst layer (CL), and (**c**) PVA on the gas diffusion layers (GDLs).

2.2. Performance Test

A fuel cell test station (HS-330s, Hephas Energy Corporation, Hsinchu, Taiwan) with an electronic load (PLZ164WA, Kikusui, Yokohama, Japan) was used for both the activation process and performance tests. Before a performance test, each MEA was activated for 12 h at the fuel cell temperature of 60 °C, anode and cathode humidifier temperatures were set at 50 °C, H_2 with a stoichiometric ratio of 1.2

was used, the minimum flow rate was 100 mL min^{-1}, and the air was prepared with a stoichiometric ratio of 3 and a minimum flow rate of 200 mL min^{-1}. During the activation process, the load was periodically changed with a minimum voltage of 0.42 V.

In the performance test, H_2 and air settings during the test were the same as during the activation. However, the fuel cell was tested in both non-humidified anode (dry anode) and humidified anode (wet anode) modes. The test conditions are shown in Table 1. During the test, voltage was measured while the current density was increased from 0 to 1.2 A cm^{-2} with a step of 0.1 A cm^{-2}.

Table 1. Temperatures of the fuel cell and anode/cathode humidifiers.

Case		Fuel Cell Temperature (°C)	Humidifier Temperature (°C) (Relative Humidity RH %)	
			Anode	Cathode
Wet	1.	60	25 (15.9)	25 (15.9)
	2.	60	30 (21.3)	30 (21.3)
	3.	60	40 (37.0)	40 (37.0)
	4.	60	50 (61.9)	50 (61.9)
	5.	60	60 (100)	60 (100)
	6.	70	60 (63.9)	60 (63.9)
	7.	70	70 (100)	70 (100)
Dry	8.	60	non-humidified	25 (15.9)
	9.	60	non-humidified	30 (21.3)
	10.	60	non-humidified	40 (37.0)
	11.	60	non-humidified	50 (61.9)
	12.	60	non-humidified	60 (100)
	13.	70	non-humidified	25 (10.2)
	14.	70	non-humidified	70 (100)

3. Results and Discussion

3.1. *Effect of PVA on MEA Performance*

Figure 2 shows the performances of MEAs without PVA operated at various humidifier temperatures. It can be seen that when both humidifier temperatures were lower than cell temperature, humidifying anode hydrogen can significantly improve the cell performance (solid lines performance is better than that of the dashed lines). However, when the humidifier temperature approached cell temperature, whether the anode was humidified or not, there was no significant effect on the cell performance, as shown in Figure 2a. Similar results can be observed at the cell temperature of 70 °C. When the cathode humidifier temperature was 60 °C, both MEAs under dry and humidified hydrogen conditions showed similar performance. However, at fully humidified conditions, in which the both the anode and cathode humidifier temperatures were the same as the cell temperature, the MEA showed a notable performance drop due to water flooding within the cell. Water existing in the CL is in both liquid and gas form. The liquid water comes with the saturated gas feed and from water absorbed by Nafion in the CL [20]. In addition to the flooding, the performance is lost by lower O_2 concentration in the saturated gas [21].

For PVA-modified MEAs, the PEMFC performance depends on the PVA content. Table 2 shows the PVA content in MEAs in this study. Performances of MEA with PVA in the anode CL (PA) under non-humidified anode operation are shown in Figure 3a–e for a fuel cell temperature of 60 °C and in Figure 3f,g for a fuel cell temperature of 70 °C. Performances of PEMFC prepared by the PA method under humidified anode operation is shown in Figures 4a–e and 4f,g for a fuel cell temperature of 70 °C and 60 °C, respectively. It is found that the PA method improved PEMFC performance only at a very low humidifier temperature, 25 °C, in both non-humidified and humidified anode cases. This result indicates that PVA is an effective additive to retain water, even at low humidity conditions. However, the addition of PVA showed a negative effect on the fuel cell performance when used at high humidify conditions (higher than the humidifier temperature of 25 °C); the performance of PEMFCs using the PA method was lower than that without PVA-MEA at other cathode humidifier temperatures (Figures 3b–e and 4b–e). It may be caused by PVA properties as a proton exchanger and electrical resistance material; adding PVA can enhance these properties of MEA. In addition, PVA might cover the catalyst active area, and the captured water may block fuel to react with the catalyst, leading to an increase in the concentration loss. Therefore, the use of the PA method could be a good way to improve the PEMFC performance when it is operated at very low humidity conditions and with a dry anode.

For a fuel cell temperature of 70 °C, low PVA loading in the anode (PA0.01 and PA0.03) showed no different performance, compared to the case of PVA-free MEA, except at a humidified anode and humidifier temperature of 70 °C. This may be because the amount of PVA caused a water balance between water transport by electro-osmosis and water diffused by back-diffusion at this operating condition. Higher operating temperature reduced ohmic loss of the fuel cell as the slope of the V-I curve at a fuel cell temperature of 70 °C and a humidifier temperature of 60 °C is less steep than that at a fuel cell temperature of 60 °C and a humidifier temperature of 60 °C. Thus, the addition of a smaller amount of PVA in the anode under this operating condition results in performance enhancement.

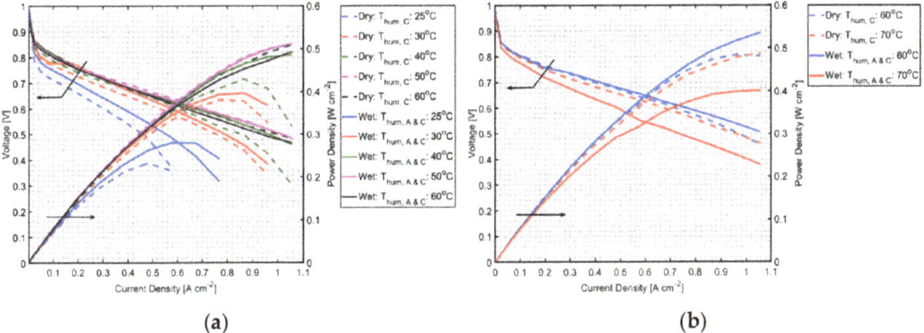

Figure 2. Performances of membrane electrode assemblies (MEAs) without PVA under non-humidified and humidified anode conditions, at a fuel cell temperature of (**a**) 60 °C and (**b**) 70 °C.

Table 2. PVA-MEA denotation.

Case	Description	PVA Loading ($\mu g\ cm^{-2}$)
PA0.01	PVA 0.01 wt.% in anode CL	0.03274
PA0.03	PVA 0.03 wt.% in anode CL	0.09823
PA0.07	PVA 0.07 wt.% in anode CL	0.22931
PA0.1	PVA 0.1 wt.% in anode CL	0.32762
PG0.03	PVA 0.5 cycles * on GDL which has PVA loading close to PA0.03 case	0.09183
PG0.1	PVA 1.75 cycles * on GDL which has PVA loading close to PA0.1 case	0.32140

*: 1 cycle of spraying PVA on GDL contains PVA of 0.18366 $\mu g\ cm^{-2}$.

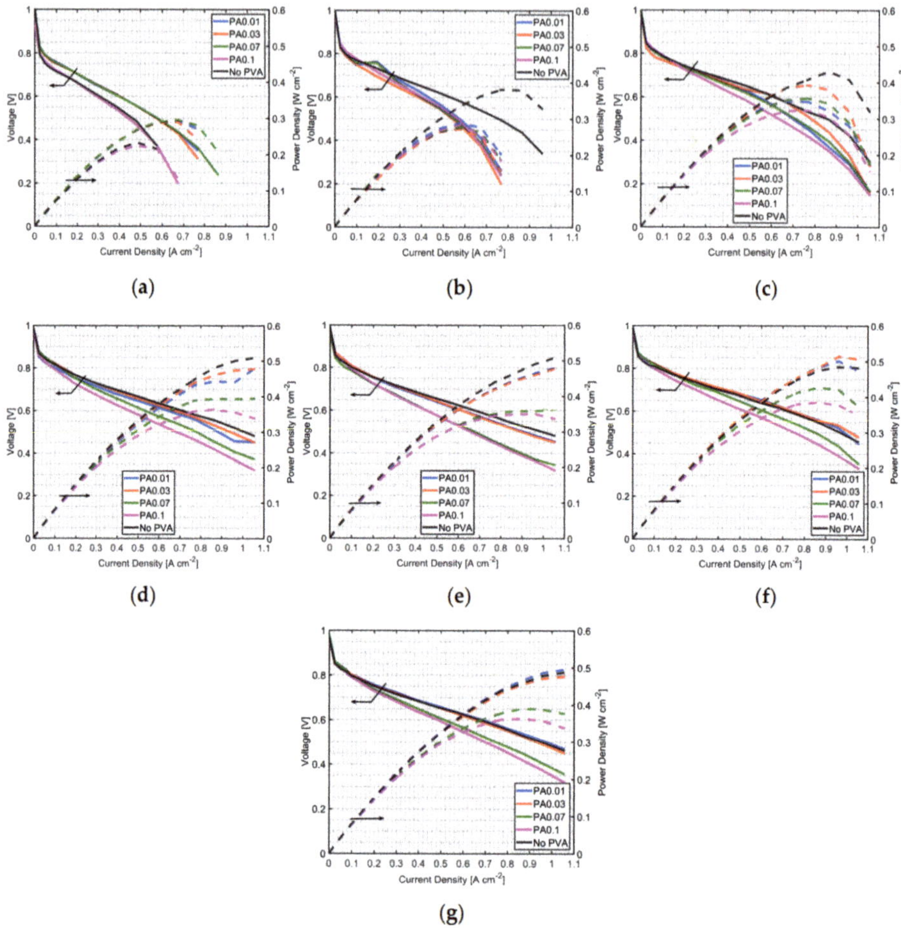

Figure 3. Performances of MEAs with the PA method at a fuel cell temperature of 60 °C, and with a non-humidified anode and cathode humidifier temperature of (**a**) 25 °C, (**b**) 30 °C, (**c**) 40°C, (**d**) 50 °C and (**e**) 60 °C; and at a fuel cell temperature of 70 °C, with a non-humidified anode, and a cathode humidifier temperature of (**f**) 60 °C and (**g**) 70 °C.

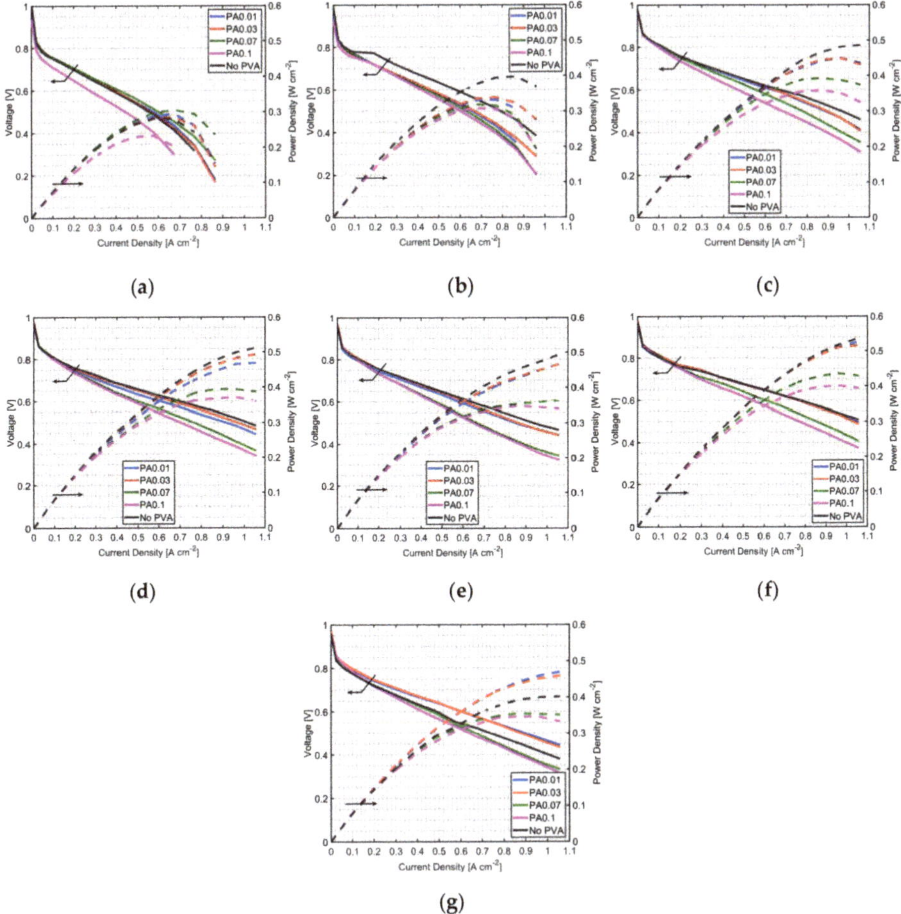

Figure 4. Performances of MEAs prepared with the PA method at a fuel cell temperature of 60 °C and the humidifier temperature of the anode and the cathode at (**a**) 25 °C, (**b**) 30 °C, (**c**) 40 °C, (**d**) 50 °C and (**e**) 60 °C; and at a fuel cell temperature of 70 °C and the humidifier temperature of the anode and the cathode at (**f**) 60 °C and (**g**) 70 °C.

3.2. Effect of PVA Location in the MEA

Because PVA causes the increase of resistance in MEAs when it is added into the anode, a new method of applying PVA, which was modified from a double layer-based technique, was proposed in this study. Instead of adding PVA into the anode catalyst, it was sprayed on the GDL as a layer between the GDL and anode CL. Due to an ultra-thin anode CL, the PVA layer could be wet by water from the back-diffusion through the membrane. However, it also could be a film between the GDL and anode CL, and it would be difficult for gas to pass through. The schematic of PVA on GDLs is shown in Figure 1c; the PVA layer is just a water reservoir, not a reaction layer. In order to compare with the PA method, PVA loading on GDLs was controlled by the number of spraying cycles, so as to have the amount of PVA close to that in the PA method. In this study, one cycle of PVA spraying on the GDL contained PVA of 0.18366 μg cm^{-2}. Table 2 shows the PVA loading of PG cases compared with PA cases. However, it is difficult to spray a PG case having the amount of PVA close to the PA0.01 case due to the very low content of PVA. Although the PG method does not affect the reaction at the anode CL, it can improve the PEMFC performance under low humidified conditions and dry anode conditions (Figures 5 and 6). PVA on GDLs can hold water from the water back-diffusion as the PEMFC performance in non-humidified anode cases (Figure 5) was changed with the moisture content in the cathode side. However, too much moisture content in the cathode side might also cause difficulty in gas transport through GDL with PVA. At low moisture content with a humidifier temperature of 25 °C, gas can go through the PVA layer, and performance is enhanced with PVA at this condition. It is found that the PG0.1 case gives better performance compared to PG0.03 due to high PVA loading at this condition; more water was adsorbed with high PVA content on the GDL and high proton exchange can be achieved at this condition. Improved PEMFC using the PG method with low PVA loading may have less resistance, but higher wettability seems to be more important to enhance the PEMFC performance. All humidified anode PG cases (Figure 6), however, show a negative effect on the performance improvement, whereas PA cases can slightly improve fuel cell performance at a humidifier temperature of 25 °C. The PVA layer is assumed to be a barrier for fuel transportation when the anode is humidified. Moisture in the anode can make the PVA swollen and results in less porosity at the surface of the GDL. Therefore, the PG method is not recommended for humidified anode application.

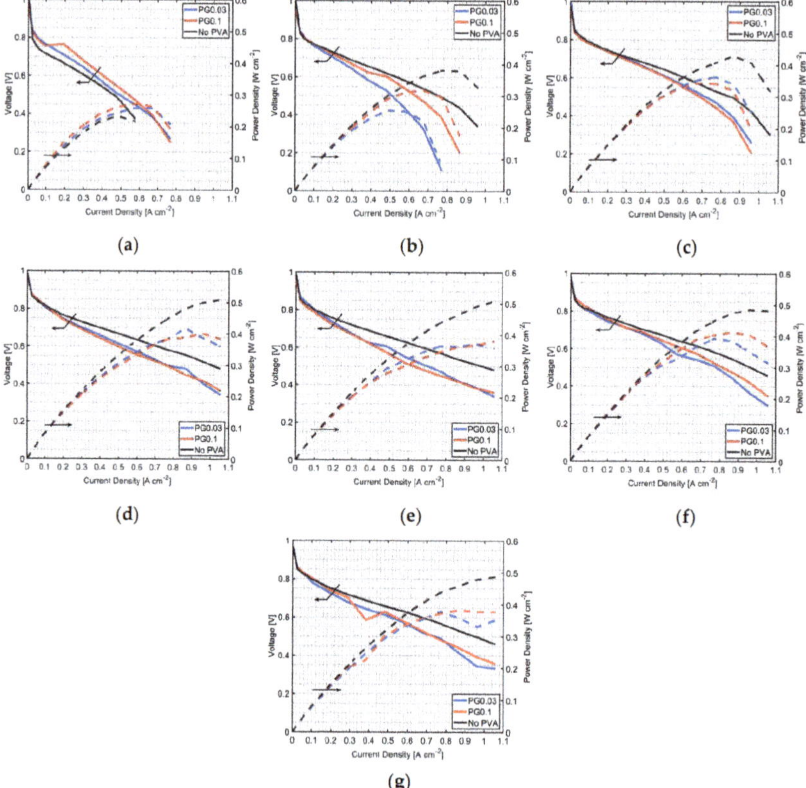

Figure 5. MEA with the PG method at a fuel cell temperature of 60 °C, with a non-humidified anode and a cathode humidifier temperature of (**a**) 25 °C, (**b**) 30 °C, (**c**) 40 °C, (**d**) 50 °C and (**e**) 60 °C; and a fuel cell temperature of 70 °C, a non-humidified anode and a cathode humidifier temperature of (**f**) 60 °C and (**g**) 70 °C.

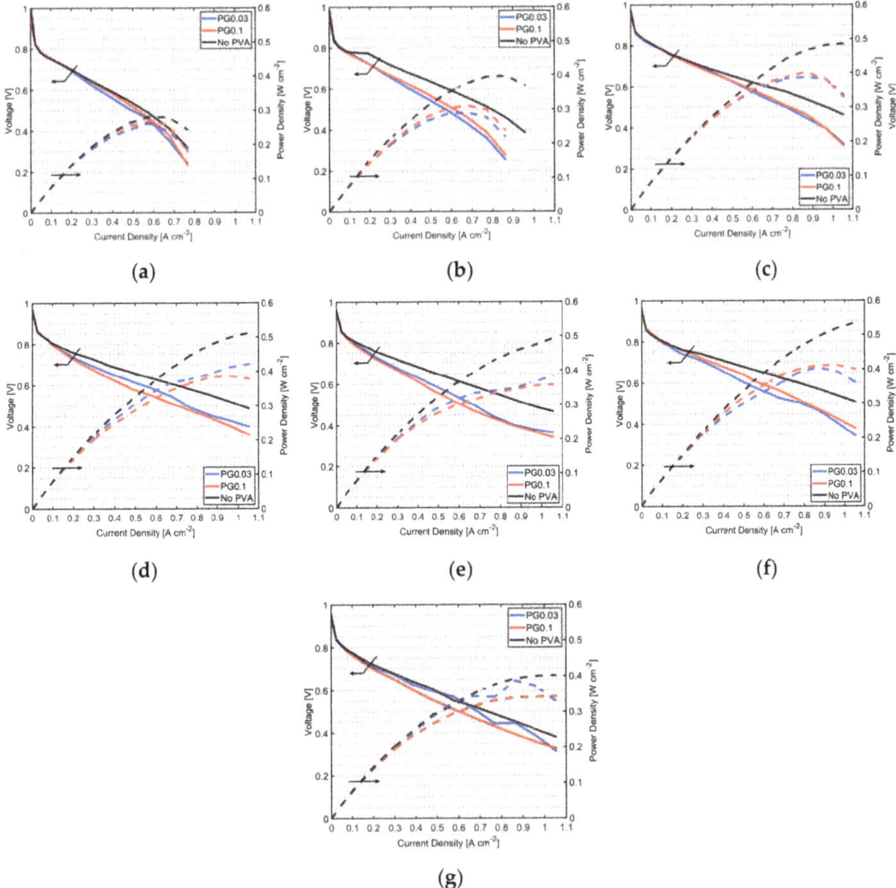

Figure 6. MEA with the PG method at a fuel cell temperature of 60 °C, and an anode and cathode humidifier temperature of (**a**) 25 °C, (**b**) 30 °C, (**c**) 40 °C, (**d**) 50 °C and (**e**) 60 °C; and a fuel cell temperature of 70 °C, and an anode and cathode humidifier temperature of (**f**) 60 °C and (**g**) 70 °C.

Figure 7a shows that with the use of a small amount of PVA into MEAs, both the PVA in the anode (PA) and PVA on GDL (PG) methods can effectively enhance the PEMFC performance when it is operated at low humidity and non-humidified anode conditions. Note that PA0.1 gives a PEMFC performance similar to using MEA without PVA, whereas PG0.1 can improve the PEMFC. It indicates that this amount of PVA using the PA method can hinder the catalyst active area, even though the membrane is wet. However, this similar amount of PVA using the PG method did not affect the catalyst active area, and thus it enhanced the performance. In comparison with PA0.03 and PG0.03, PA0.03 provided better performance if the current density was higher than 0.3 A cm^{-2}. It might be because the PVA content in the anode was able to keep the membrane hydrated, and dominated the effect of PVA on the catalyst surface area. Thus, PVA location in the MEA plays an important role in cell improvement. PVA was preferred to be a bit distant from the membrane when higher PVA loading was used, as shown in the PA0.1-PG0.1 cases. For lower PVA loadings like the PA0.03-PG0.03 cases, PVA was instead placed inside the anode CL close to the membrane. The PEMFC improvement by PVA at a cell voltage of 0.6 V, a non-humidified anode, and a cathode humidifier temperature of 25 °C is shown in Figure 7b. Adding a small amount of PVA into the MEA can improve the PEMFC current density by around 30% at selected operating conditions.

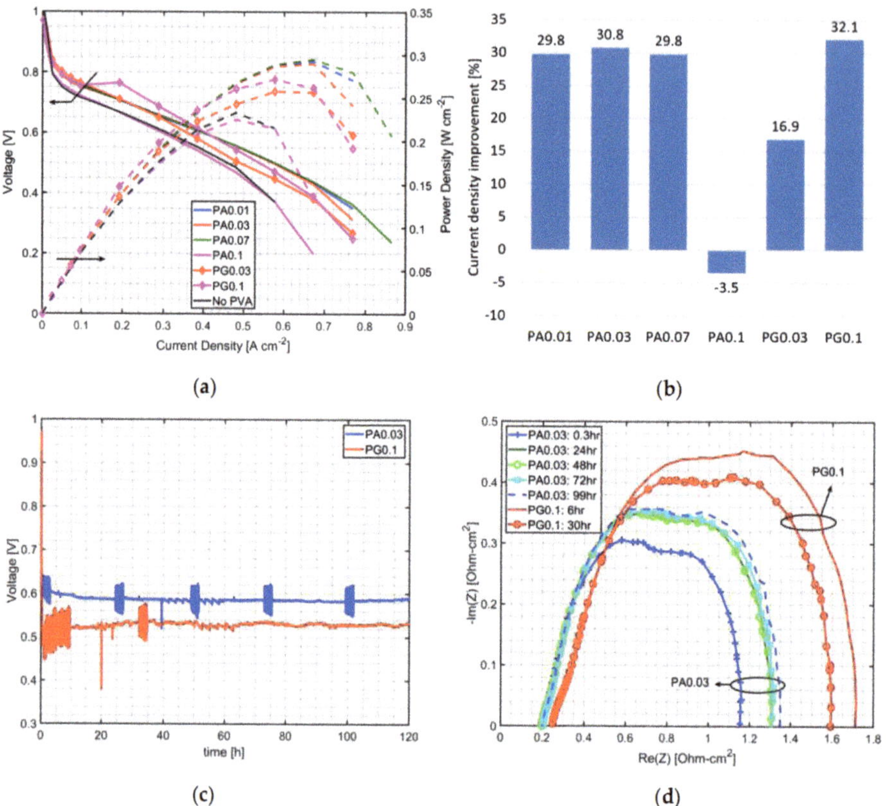

Figure 7. (**a**) Comparison of PEMFCs using PVA-MEA prepared by the PA and PG methods at a non-humidified anode and cathode humidifier temperature of 25 °C, (**b**) percent improvement of PVA-MEA from MEA without PVA at the non-humidified anode, with a cathode humidifier temperature of 25 °C and a voltage of 0.6 V, (**c**) the durability test and (**d**) the impedance curve of PA0.03 and PG0.1 at a constant current density of 0.4 A cm^{-2}, with a non-humidified anode and a cathode humidifier temperature of 25 °C.

3.3. Durability Test

For a durability test, both methods are examined by comparing the voltage variation of those MEAs at the constant current density of 0.4 A cm^{-2}, which is the result of short-term test conditions to achieve the target voltage of 0.6 V (Figure 7a), and constant feed flow rates for 120 h. However, no durability test of the MEA without PVA is presented in Figure 7c,d because the performance of the MEA without PVA showed a lower performance than the PA0.03 and PG0.1 cases. According to Figure 7a, the voltage of the MEA without PVA can start from 0.55V when the current density is at 0.4 A cm^{-2}, and a quick performance drop will occur under dry operating conditions; therefore, no long term test data can be provided for the MEA without PVA with the target voltage of 0.6 V and a given constant current density of 0.4 A cm^{-2}. PA0.03 and PG0.1 are selected as representative of the PA and PG methods, respectively, due to highest performance improvement in each method. Although PG0.1 showed a slightly higher performance than PA0.03 (Figure 7b), greater voltage drop (about 13.3%) is found in the

PG case in a long term test (Figure 7c); whereas voltage drops of about 3.4% were found in the PA case. Impedance curves of the PA and PG methods (Figure 7d) were measured at the current density of 0.4 A cm^{-2} by electrochemical impedance spectroscopy with frequency sweeping from 10 mHz to 10 kHz. It shows that the PA method has smaller resistance, both membrane resistance (ohmic loss) and polarization resistance (activation losses and mass transport losses), than the PG method. Higher resistance in the PG case might be because higher PVA loading was used in the PG case, and fuel had more difficulty passing through a swollen PVA layer when water was accumulated with time; this is observed from the polarization resistance increase with time (Figure 7d). The periodic fluctuation in voltage was due to the measurement of electrochemical impedance, in which the frequency was changed. In comparison with PVA-MEA using a similar fabrication technique (spraying method), the achieved maximum power density proposed by Liang et al. [19] is higher than the PVA-MEA prepared by both the PA and PG methods. The lower achievement could be due to differences in operating conditions and material compositions. However, the proposed methods in this study can be used in the lower humidity condition (RH of 15.9% and a non-humidified anode). Roh et al. [12] tested the oxygen-group functionalized carbon supported dual catalyst layer of PEMFC under RH of 26%. The results were superior to the proposed method in this study. However, it is more complicated and costly for PEMFC fabrication.

4. Conclusions

This study is aimed at improving the performance of PEMFCs operated at an ambient environment with low humidity, using PVA as an additive. An ultrasonic spraying method was employed to coat a CL on a Nafion 211 membrane. PVA was applied to the MEA by two methods: PVA in the anode and PVA on GDLs (a newly proposed method). PVA was added in the anode during the catalyst ink preparation procedure for PVA in the anode (PA) case. For PVA on the GDL (PG), PVA was sprayed on the anode GDL with the same coating machine. Although the use of PVA can increase the wettability of the MEA's PEMFC, low PVA content was required to minimize the cell resistance caused by the PVA. Both PA and PG methods can enhance the PEMFC performance only at very low humidity condition, especially under non-humidified anode condition. The current density of the PVA-MEA prepared by the PA and PG methods could be improved by approximately 30%, at the operating voltage of 0.6 V, and non-humidified anode and cathode humidifier temperature of 25 °C. However, the durability test revealed that the PA method is more durable than the PG method, due to a smaller voltage drop.

Author Contributions: Conceptualization, Y.-S.C. and A.A.; methodology, data curation, and writing—original draft preparation, P.J.; writing—review and editing, Y.-S.C. and A.A. All authors have read and agreed to the published version of the manuscript.

Funding: This research was funded by the Thailand Program Management Unit-B (PMU-B), Thailand; Advanced Institute of Manufacturing with High-tech Innovations (AIM-HI) from The Featured Areas Research Center Program within the framework of the Higher Education Sprout Project by the Ministry of Education (MOE) and the Ministry of Science and Technology under project No. 106-2221-E-194-033-MY3 in Taiwan. P.J. would also like to acknowledge the postdoctoral fellowship, from the Ratchadapisek Sompoch Endowment Fund (Chulalongkorn University).

Conflicts of Interest: The authors declare no conflict of interest.

References

1. Oh, T.H. Conceptual design of small unmanned aerial vehicle with proton exchange membrane fuel cell system for long endurance mission. *Energy Convers. Manag.* **2018**, *176*, 349–356. [CrossRef]
2. Teixeira, F.C.; de Sá, A.I.; Teixeira, A.P.S.; Rangel, C.M. Nafion phosphonic acid composite membranes for proton exchange membranes fuel cells. *Appl. Surf. Sci.* **2019**, *487*, 889–897. [CrossRef]
3. Yang, T.-H.; Yoon, Y.-G.; Kim, C.-S.; Kwak, S.-H.; Yoon, K.-H. A novel preparation method for a self-humidifying polymer electrolyte membrane. *J. Power Sources* **2002**, *106*, 328–332. [CrossRef]

4. Choi, I.; Lee, H.; Lee, K.G.; Ahn, S.H.; Lee, S.J.; Kim, H.-J.; Lee, H.-N.; Kwon, O.J. Characterization of self-humidifying ability of SiO_2-supported Pt catalyst under low humidity in PEMFC. *Appl. Catal. B* **2015**, *168–169*, 220–227. [CrossRef]
5. Tian, J.; Gao, P.; Zhang, Z.; Luo, W.; Shan, Z. Preparation and performance evaluation of a Nafion-TiO_2 composite membrane for PEMFCs. *Int. J. Hydrogen Energy* **2008**, *33*, 5686–5690. [CrossRef]
6. Mollá, S.; Compañ, V.; Gimenez, E.; Blazquez, A.; Urdanpilleta, I. Novel ultrathin composite membranes of Nafion/PVA for PEMFCs. *Int. J. Hydrogen Energy* **2011**, *36*, 9886–9895. [CrossRef]
7. Ahn, M.-K.; Lee, S.-B.; Min, C.-M.; Yu, Y.-G.; Jang, J.; Gim, M.-Y.; Lee, J.-S. Enhanced proton conductivity at low humidity of proton exchange membranes with triazole moieties in the side chains. *J. Membr. Sci.* **2017**, *523*, 480–486. [CrossRef]
8. Kim, K.; Heo, P.; Ko, T.; Kim, K.-H.; Kim, S.-K.; Pak, C.; Lee, J.-C. Poly (arlyene ether sulfone) based semi-interpenetrating polymer network membranes containing cross-linked poly (vinyl phosphonic acid) chains for fuel cell applications at high temperature and low humidity conditions. *J. Power Sources* **2015**, *293*, 539–547. [CrossRef]
9. Patel, H.A.; Mansor, N.; Gadipelli, S.; Brett, D.J.L.; Guo, Z. Superacidity in Nafion/MOF hybrid membranes retains water at low humidity to enhance proton conduction for fuel cells. *ACS Appl. Mater. Interfaces* **2016**, *8*, 30687–30691. [CrossRef] [PubMed]
10. Yin, Y.; Liu, J.; Chang, Y.; Zhu, Y.; Xie, X.; Qin, Y.; Zhang, J.; Jiao, K.; Du, Q.; Guiver, M.D. Design of Pt-C/Fe-N-S-C cathode dual catalyst layers for proton exchange membrane fuel cells under low humidity. *Electrochim. Acta* **2019**, *296*, 450–457. [CrossRef]
11. Guo, X.; Shao, Z.; Xiao, Y.; Zeng, Y.; Liu, S.; Wang, X.; Yi, B. Improvement of the proton exchange membrane fuel cell (PEMFC) performance at low-humidity conditions by exposing anode in ultraviolet light. *Electrochem. Commun.* **2014**, *44*, 16–18. [CrossRef]
12. Roh, C.-W.; Choi, J.; Lee, H. Hydrophilic-hydrophobic dual catalyst layers for proton exchange membrane fuel cells under low humidity. *Electrochem. Commun.* **2018**, *97*, 105–109. [CrossRef]
13. Su, H.; Xu, L.; Zhu, H.; Wu, Y.; Yang, L.; Liao, S.; Song, H.; Liang, Z.; Birss, V. Self-humidification of a PEM fuel cell using a novel Pt/SiO_2/C anode catalyst. *Int. J. Hydrogen Energy* **2010**, *35*, 7874–7880. [CrossRef]
14. Han, M.; Chan, S.H.; Jiang, S.P. Investigation of self-humidifying anode in polymer electrolyte fuel cells. *Int. J. Hydrogen Energy* **2007**, *32*, 385–391. [CrossRef]
15. Leimin, X.; Shijun, L.; Lijun, Y.; Zhenxing, L. Investigation of a novel catalyst coated membrane method to prepare low-platinum-loading membrane electrode assemblies for PEMFCs. *Fuel Cells* **2009**, *9*, 101–105. [CrossRef]
16. Chen, F.; Liu, P. High electrically conductive polyaniline/partially phosphorylated poly (vinyl alcohol) composite films via aqueous dispersions. *Macromol Res* **2011**, *19*, 883. [CrossRef]
17. El-Toony, M.M.; Abdel-Hady, E.E.; El-Kelsh, N.A. Casting of poly hydroxybutarate/poly (vinyl alcohol) membranes for proton exchange fuel cells. *Electrochim. Acta* **2014**, *150*, 290–297. [CrossRef]
18. Attaran, A.M.; Javanbakht, M.; Hooshyari, K.; Enhessari, M. New proton conducting nanocomposite membranes based on poly vinyl alcohol/poly vinyl pyrrolidone/$BaZrO_3$ for proton exchange membrane fuel cells. *Solid State Ion.* **2015**, *269*, 98–105. [CrossRef]
19. Liang, H.; Zheng, L.; Liao, S. Self-humidifying membrane electrode assembly prepared by adding PVA as hygroscopic agent in anode catalyst layer. *Int. J. Hydrogen Energy* **2012**, *37*, 12860–12867. [CrossRef]
20. Jiao, K.; Li, X. Water transport in polymer electrolyte membrane fuel cells. *Prog. Energy Combust. Sci.* **2011**, *37*, 221–291. [CrossRef]
21. Zhang, G.; Jiao, K. Multi-phase models for water and thermal management of proton exchange membrane fuel cell: A review. *J. Power Sources* **2018**, *391*, 120–133. [CrossRef]

Article

Ionic Liquid-Incorporated Zn-Ion Conducting Polymer Electrolyte Membranes

Jianghe Liu [1], Sultan Ahmed [1], Zeba Khanam [1], Ting Wang [2] and Shenhua Song [1,*]

[1] Shenzhen Key Laboratory of Advanced Materials, School of Materials Science and Engineering, Harbin Institute of Technology, Shenzhen 518055, China; liujianghe@stu.hit.edu.cn (J.L.); usmanisultan@gmail.com (S.A.); zbkhnm@gmail.com (Z.K.)

[2] Guangdong Provincial Key Laboratory of Electronic Functional Materials and Devices, Huizhou University, Huizhou 516001, China; twang-hawk@foxmail.com

* Correspondence: shsong@hit.edu.cn; Tel.: +86-755-26033465

Received: 15 July 2020; Accepted: 31 July 2020; Published: 6 August 2020

Abstract: In this study, novel ionic liquid-incorporated Zn-ion conducting polymer electrolyte membranes containing polymer matrix poly (vinylidene fluoride-hexafluoropropylene) (PVdF-HFP) and 1-ethyl-3-methylimidazolium trifluoromethanesulfonate (EMITf), along with zinc trifluoromethanesulfonate $Zn(Tf)_2$, are prepared and investigated. It is ascertained that the optimal membrane ILPE-Zn-4 (the mass ratio of EMITf:$Zn(Tf)_2$:PVDF-HFP is 0.4:0.4:1), with abundant nanopores, exhibits a high amorphousness. At room temperature, the optimized electrolyte membrane offers a good value of ionic conductivity (~1.44×10^{-4} S cm^{-1}), with a wide electrochemical stability window (~4.14 V). Moreover, the electrolyte membrane can sustain a high thermal decomposition temperature (~305 °C), and thus its mechanical performance is sufficient for practical applications. Accordingly, the ionic liquid-incorporated Zn-ion conducting polymer electrolyte could be a potential candidate for Zn-based energy storage applications.

Keywords: polymer electrolyte; Zn-ion conducting; ionic liquid; ionic conductivity

1. Introduction

Owing to the development of energy storage technology around the globe, mobile electronics have seen a rapid growth. To make our lives more convenient, various electronic products like smartphones, laptops, digital cameras, electric vehicles, etc. have been integrated into our daily life. To date, Li-ion batteries have been widely applied in electronic equipment [1]. However, the uneven geo-distribution of elemental Li and high cost seriously restrict the development of Li-ion batteries [2,3]. Therefore, more abundant and lower-cost metal ions (monovalent cations: Na^+, K^+, multivalent cations: Mg^{2+}, Zn^{2+}, and Al^{3+}) are preferred for energy storage systems [4]. Among them, Zn is present in abundance on earth. Furthermore, it has a low cost and a relatively higher volumetric capacity of ~5855 mAh cm^{-3} in comparison to Na (~1129 mAh cm^{-3}), Li (~2061 mAh cm^{-3}), and Mg (~3834 mAh cm^{-3}) [5]. Inspired by these virtues, the Zn-based energy storage system has been drawing considerable attention. Zn-ion batteries usually use Manganese Dioxides, Vanadium Compounds, Prussian Blue Analogs as cathodes and metal Zn plates as anodes [6–8]. The aqueous electrolyte solutions of $ZnSO_4$, $Zn(CH_3COO)_2$, $Zn(CF_3SO_3)_2$ have been widely studied in Zn-ion batteries [9–11]. However, the work voltage window of aqueous electrolytes is narrow (~1.2 V). Though organic electrolytes offer a high ionic conductivity and wide voltage window, they suffer from other problems like leakage, flammability, corrosiveness, and toxicity [12]. Due to their excellent properties, like a high thermal stability, relatively higher electrochemical stability, high flexibility, and safety, ionic conductive polymer electrolytes have emerged as potential candidates for use in the electrochemical energy storage field [13,14].

In general, on account of the different compositions of the polymer electrolyte system, ionic conductive polymer electrolytes have been divided into two classes. These are solid polymer electrolytes (SPEs) and gel polymer electrolytes (GPEs). For SPEs, the ionic salts are complexed with flexible polymer chains, and the transportation of ions depends on the motion of polymer segments [15]. GPEs are formed by ionic salts and plasticizer mixed into polymer matrices. Due to their high room-temperature ionic conductivity and contact with electrodes with a good compatibility, GPEs have attracted considerable attention [16]. Traditional plasticizers, having a low chemical and electrochemical stability and inflammability, include propylene carbonate (PC), ethylene carbonate (EC), ethyl methyl carbonate (EMC), dimethyl carbonate (DMC), etc. [17,18]. Ionic liquid with excellent electrochemical properties can be a good substitute for traditional plasticizers. For this reason, the imidazolium-based ionic liquid (1-ethyl-3-methylimidazolium trifluoromethanesulfonate, EMITf) was chosen as a plasticizer in the present study. This ionic liquid exhibits a lower viscosity and higher thermal stability [17,19]. Moreover, ionic liquid can also provide a lot of free ions so as to promote the ionic conductivity of polymer electrolytes [20]. In addition to plasticizers, GPEs contain various functional polymer hosts. Generally, these are poly (vinyl alcohol) (PVA), poly(acrylonitrile) (PAN), poly (methyl methacrylate) (PMMA), and poly (vinylidene fluoride-hexafluoropropylene) (PVdF-HFP) [14,21–24]. Among them, PVdF-HFP with a low crystallinity and high dielectric constant is chosen as the polymer host [25–27]. PVdF-HFP consists of both amorphous and crystalline structures. The amorphous region is beneficial to ion transport, while the crystalline structure provides a better mechanical property [28–30]. Furthermore, PVdF-HFP exhibits sound chemical and thermal stabilities due to its strong C-F bonds [31,32]. Therefore, in recent years, the ionic liquid-doped PVdF-HFP polymer electrolyte has been studied intensively. Guo et al. [23] developed a polymer electrolyte based on the PVDF-HFP/LiTFSI/SiO_2/EMITFSI system, and an optimum electrolyte with an ionic conductivity of 0.74 mS cm $^{-1}$ at 25 °C was applied in a Li/$LiFePO_4$ battery. Such a battery showed an excellent cycling stability. Singh et al. [17] fabricated a flexible GPE containing PVdF-HFP and EMIMFSI, along with LiTFSI. The GPE showed the highest ionic conductivity, at ~3.8×10^{-4} S cm^{-1} at 25 °C. Kumar et al. [33] reported an electrolyte membrane based on the PVdF-HFP/$NaCF_3SO_3$/EMITf system, and the electrolyte membrane offered an ionic conductivity of ~ 5.7×10^{-3} S cm^{-1} at room temperature. Tang et al. [34] revealed that a GPE film of PVdF-HFP:$Mg(Tf)_2$ (9:1) doped with 40 wt.% EMITf exhibited a high room-temperature ionic conductivity of ~4.63×10^{-3} S cm^{-1} with a wide electrochemical stability window of ~4.8 V. Thus, it was inferred that the ionic liquid-doped PVdF-HFP polymer electrolyte exhibits excellent electrochemical properties.

In the present work, we have prepared ionic liquid-incorporated Zn-ion conducting polymer electrolytes, and studied their structural, electrical, electrochemical, thermal, and mechanical properties. A detailed study on solid polymer electrolytes based on the PVdF-HFP/$Zn(Tf)_2$ system was conducted in our previous work [35]. The SPE exhibited a good electrochemical performance when the mass ratio of $Zn(Tf)_2$ to PVdF-HFP was 0.4. Hence, we chose the optimal composition to be the host system in the present work. The electrochemical properties of the polymer electrolyte were enhanced by doping ionic liquid. The ionic liquid-incorporated polymer electrolytes were investigated by means of structural, electrical, electrochemical, thermal, and mechanical analyses.

2. Materials and Methods

The following materials were used in the preparation of ionic liquid-doped polymer electrolytes and supercapacitors. Poly (vinylidene fluoride-co-hexafluoropropylene) (PVdF-HFP, MW ~40,000), Zinc trifluoromethanesulfonate ($Zn(Tf)_2$, purity ~99.6%), and tetrahydrofuran (THF, purity ~99.5%) were acquired from Sigma Aldrich. Ionic liquid 1-ethyl-3-methylimidazolium trifluoromethanesulfonate (EMITf, purity ~99.8%) was obtained from Aladdin. The chemicals were used as received. Both the ionic liquid EMITf and salt $Zn(Tf)_2$ were vacuum-dried at 100 °C for 2 h prior to use to remove the moisture.

According to our previous work [30], the solid polymer electrolyte exhibits the best electrochemical properties when the mass ratio of $Zn(Tf)_2$ to PVdF-HFP is 0.4, and this composition was chosen as the host system for the preparation of ionic liquid-incorporated Zn-ion polymer electrolyte (ILPE) membranes. Moreover, ILPE membranes with different contents of ionic liquid (the mass ratio of EMITf to PVdF-HFP being 0, 0.1, 0.2, 0.3, 0.4, and 0.5, marked as SPE-Zn, ILPE-Zn-1, ILPE-Zn-2, ILPE-Zn-3, ILPE-Zn-4, and ILPE-Zn-5, respectively) were prepared via a solution cast method. First, the host polymer, PVdF-HFP, was dissolved in THF with the help of magnetic stirring at 50 °C. Subsequently, $Zn(Tf)_2$ was added to the solution in a mass ratio of $Zn(Tf)_2$ to PVdF-HFP of 0.4, followed by continuous stirring at 50 °C for 2 h. Afterwards, different contents of the ionic liquid EMITf were added, and the mixture was continuously stirred for 18 h to obtain a homogenous viscous solution. Thereafter, the resulting homogeneous solution was poured into different Petri dishes, followed by the evaporation of THF at 35 °C for 10 h. Finally, the obtained free-standing ILPE membranes were stored in a vacuum oven at 40 °C for 4 h, so that the THF could be completely removed. The macroscopic morphology of the obtained electrolyte membrane is shown in Figure 1. It may be noted that the SPE-Zn membrane is uniformly transparent and impurity-free. The membrane becomes translucent with the addition of EMITf (see Figure 1b,c). All the membranes display a good flexibility. The thickness of the membranes was measured to be ~145 μm via a spiral micrometer gauge. The as-prepared electrolyte membranes were stored in a glove box filled with Ar atmosphere.

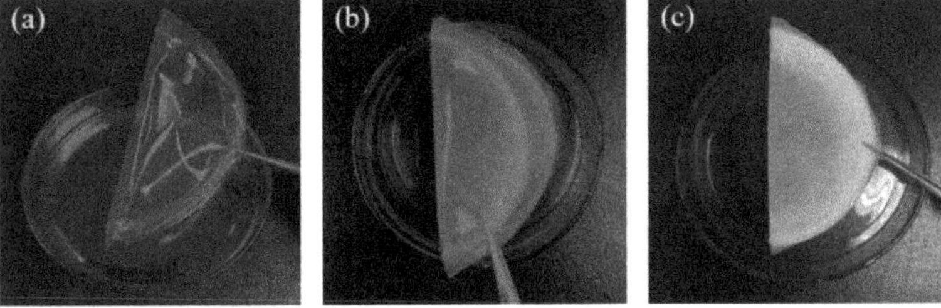

Figure 1. Macroscopic morphologies of the (**a**) SPE-Zn membrane, (**b**) ILPE-Zn-4 membrane, and (**c**) ILPE-Zn-5 membrane.

The crystal structure and crystallinity of the polymer electrolyte membranes were examined using X-ray diffraction (XRD, D/max 2500 PC, Rigaku, Japan) in the 2θ range of 5–65° with a 2°/min scan rate. Scanning electron microscopy (SEM, Hitachi S-4700, Tokyo, Japan) was used to observe the morphological characteristics of the electrolyte membranes. The thermal stability of the ILPE membranes was studied by a thermal analyzer (STA 449F3 Jupiter, Selb, Germany) under an Ar atmosphere at a rate of 10°/min between 30 and 600 °C. Using a universal testing machine (CMT7504, Ningbo, China), the mechanical performance of the electrolyte membranes was evaluated. The sample size was: thickness (0.14 mm) × width (10 mm) × gauge length (100 mm). The ionic conductivity analysis for the electrolyte membranes was carried out using a frequency analyzer (PSM 1735, Newton, UK) in the range of 1 Hz to 1 MHz. The electrochemical stability window (ESW) and ionic transference number were evaluated by cyclic voltammetry (CV) and DC polarization method, separately, by means of a CHI760D electrochemical workstation. The scan range was −3 to 3 V with a scan rate of 10 mV s^{-1} for the CV measurements, and the applied voltage was 0.5 V for the determination of the ionic transference number with DC polarization. As for the above electrochemical measurements, the ILPE membranes were assembled in a symmetric sandwich structure cell with stainless steel electrodes (SS/ILPE membrane/SS), in which the electrolyte-electrode contact area was 1.96 cm^2. Finally, to assess the practicality of an optimal electrolyte membrane as an ionic conductor, it was connected into the circuit.

3. Results and Discussion

XRD was used to analyze the structure and crystallinity of PVdF-HFP based gel electrolytes. The analyzed results are shown in Figure 2. The pure PVDF-HFP consists of crystalline regions with amorphous domains and presents a typical semi-crystalline characteristic. However, with the addition of $Zn(Tf)_2$ to pristine PVdF-HFP, there is a dramatic change in the structure of PVdF-HFP. The intensity of the peaks decreases, and their width increases, thereby expanding the amorphous region of the host polymer [35]. It may be noted that, with the addition of ionic liquid, there is a suppression in the peak intensity. This indicates that the ionic coordination between ionic liquid and polymeric macromolecules with $Zn(Tf)_2$ may interrupt the structure of the polymer and weaken the intermolecular interaction of the polymer chains, thereby reducing the degree of crystallinity and thus enhancing the amorphous behavior [34,36]. However, when the addition of ionic liquid is too large, the crystallinity of the electrolyte membranes increases, rather than further decreasing. This may be attributed to the re-association of excess Tf^- anions with Zn^{2+} cations, weakening the interaction between ions and polymer segments [37]. It may be noted that, for ILPE-Zn-4, the peak at 2θ ~26.6° is completely suppressed and that, accordingly, the amorphousness of the host polymer has reached its maximum. Moreover, the mobile charge carriers can be transported speedily in amorphous regions in the polymer electrolyte system because they have more free spaces, hereby facilitating the migration of ions [38]. Therefore, it is believed that the ILPE-Zn-4 with larger amorphous regions should be favorable for ion movement.

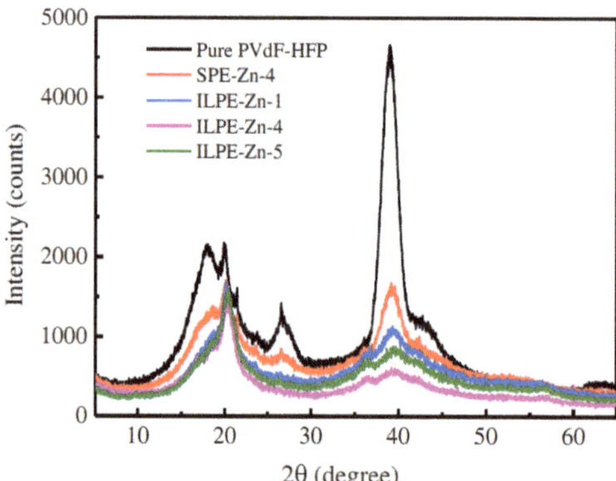

Figure 2. XRD patterns of ILPE membranes with different EMITf contents.

Figure 3 displays the SEM images of the ILPE membranes. The surface of the solid polymer electrolyte SPE-Zn can be seen as being wrinkled, with barely visible micropores (see Figure 3a). However, the cross-section morphology of SPE-Zn implies that there are a number of nanopores inside with network structures. For ILPE-Zn-4 (see Figure 3b,d), the addition of ionic liquid EMITf improves the surface morphology in terms of nanopores. It may be noted that the nanopores are interconnected with each other, which may be attributed to the interaction between polymeric macromolecules and ionic liquid, in addition to the evaporation of the THF solvent [39].

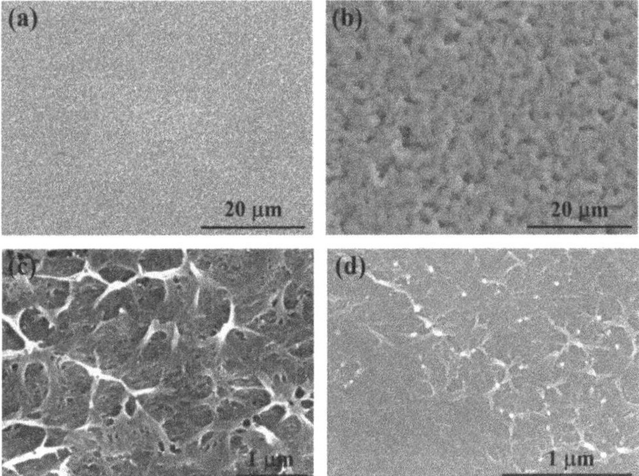

Figure 3. SEM images of (**a**) SPE-Zn, (**b**) ILPE-Zn-4, (**c**) SPE-Zn cross-section, and (**d**) ILPE-Zn-4 cross-section.

There is a significant change in the internal structure of the polymer electrolyte with the addition of ionic liquid. The pores become more abundant, with fine sizes. In general, this is beneficial for ion transportation, as the small interconnected nanopores' structure and high specific surface areas provide a continuous pathway for ion transportation [34,40,41]. As a consequence, it is speculated that the introduction of ionic liquid to the polymer electrolyte may enhance its ionic conductivity.

The room-temperature ionic conductivity of polymer electrolytes is one of the vital parameters to consider. AC impedance technique was used to analyze the ion conduction behavior of polymer electrolytes. Figure 4a shows the Nyquist plots of gel electrolyte membranes measured at ambient temperature, and the inset shows their equivalent circuit. It is clear that the plots consist of a quasi-semicircle with a spike line. The quasi-semicircle in the high frequency range is present as a result of the parallel combination of the resistor and constant phase element. The constant phase element may be regarded as a non-ideal capacitor. The spike line located in the low frequency range indicates the formation of a double-layer capacitor at the interface between the electrode and electrolyte [42]. It is noted that the dip angle between the spike line and the real axis is smaller than 90° and arises from the roughness of the electrolyte-electrode interface [43]. The bulk resistance (R_b) of the electrolyte membrane can be determined from the intercept of the semicircle on the real axis. Clearly, the bulk resistance of the polymer electrolyte decreases with the addition of EMITf and exhibits a minimum magnitude when the mass ratio of EMITf to PVdF-HFP is 0.4. Moreover, the high frequency quasi-semicircle almost disappears for a mass ratio of 0.4, indicating that the ILPE-Zn-4 electrolyte prevails in the resistive nature and that its capacitive character is absent [44]. In such a case, the bulk resistance (R_b) is usually obtained from the interception of the inclination spike on the real axis [45,46]. The ionic conductivity of ILPE membranes is given by Formula (1) [47], and the results are listed in Table 1.

$$\sigma = \frac{d}{R_b S} \tag{1}$$

where R_b is the bulk resistance (Ω), d represents the membrane thickness (cm), and S is the electrolyte-electrode contact area (cm^2).

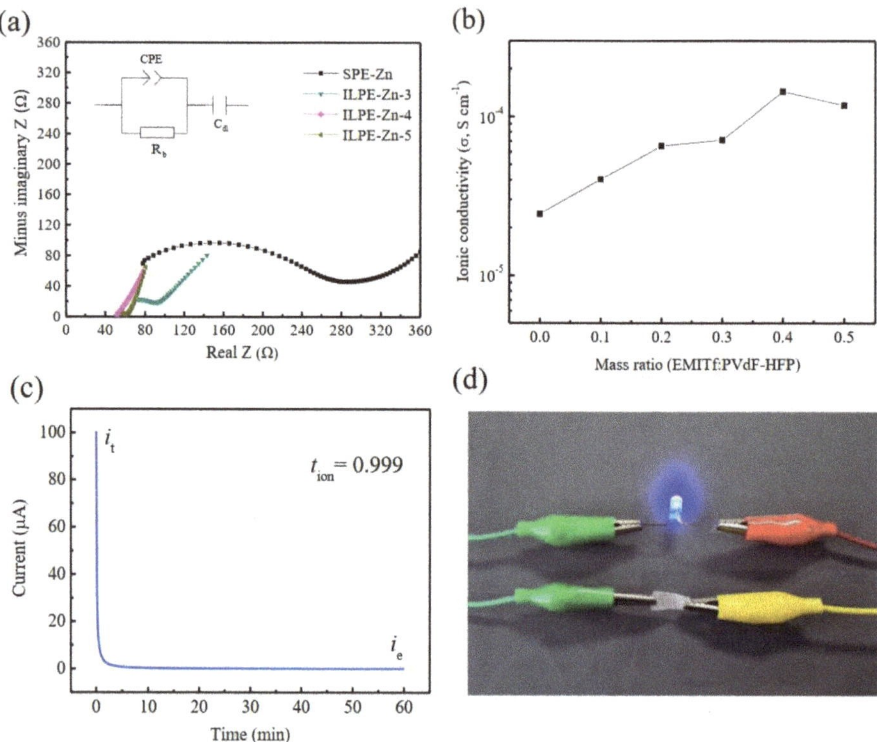

Figure 4. (**a**) Nyquist plots of the ILPE membranes with different EMITf contents measured at ambient temperature and corresponding equivalent circuit (CPE = constant phase element, R_b = resistor, and C_{dl} = double-layer capacitor). (**b**) Variation of ambient temperature ionic conductivity with the mass ratio of EMTIf to PVdF-HFP. (**c**) DC polarization curve of ILPE-Zn-4 membrane under 0.5 V at room temperature. (**d**) The ILPE-Zn-4 membrane used as an ionic conductor to successfully connect the LED circuit.

Table 1. Ionic conductivities of polymer electrolyte membranes at room temperature.

Samples	Ionic Conductivities (S cm^{-1})
SPE-Zn	2.44×10^{-5}
ILPE-Zn-1	4.03×10^{-5}
ILPE-Zn-2	6.56×10^{-5}
ILPE-Zn-3	7.15×10^{-5}
ILPE-Zn-4	1.44×10^{-4}
ILPE-Zn-5	1.18×10^{-4}

Figure 4b depicts the variation of the ambient temperature ionic conductivity with the mass ratio of EMITf to PVdF-HFP. It may be noted that the ionic conductivity of the SPE-Zn membrane without ionic liquids is low, being just ~2.44×10^{-5} S cm^{-1}. However, the ionic conductivity of the ILPE membrane is improved by adding ionic liquid EMITf into the system, and it shows a high value of approximately 1.44×10^{-4} S cm^{-1} when the mass ratio is 0.4. With the further addition of EMITf, the ionic conductivity decreases. Accordingly, in terms of ionic conductivity, the ILPE-Zn-4 is the best composition. In general, the ionic conductivity (σ) depends on three factors: the concentration (n_i), mobility (q_i), and charge (u_i) of mobile ions. The relationship between them can be described by [48]:

$$\sigma = \sum n_i q_i u_i \quad (2)$$

Combined with the above structural analysis, it is believed that the enhancement of the ionic conductivity may be ascribed to both the improvement of the amorphous phase and the increase of mobile ions, caused by the addition of EMITf [17]. However, the addition of excessive ionic liquid would make the ions aggregate, thereby reducing the density of the mobile charge carriers and hindering their movement [39]. Consequently, the ionic conductivity decreases.

In order to confirm that the prepared polymer electrolyte is highly ionically conductive, which is vital for its application, the ionic transference number was determined by DC polarization at room temperature. Figure 4c shows the DC polarization current versus time for the ILPE-Zn-4 polymer electrolyte membrane. Generally, the initial current (i_t) is attributed to the transportation of both ions and electrons. It can be observed that the current decreases dramatically with an increasing polarization time until reaching a steady-state value. This is primarily due to the blockage of ions by the electrodes. Therefore, the final steady-state current (i_e) is merely contributed by electrons. The ionic transference number is calculated by [49]:

$$t_{ion} = (i_t - i_e)/i_t \quad (3)$$

The ionic transference number of the ILPE-Zn-4 membrane is obtained as being ~0.999 ± 0.0003. The ultrahigh ionic transference number demonstrates that the conductivity of the polymer electrolyte membrane is mainly caused by ionic species, while the electronic contribution is too small to be considered. Therefore, the present polymer electrolyte membrane is almost 100% ionically conductive. Moreover, the ultrahigh ionic transference number ~0.999 is superior to the ionic transference number reported in literatures, such as the PVdF-HFP/$LiBF_4$/$EMIMBF_4$ electrolyte (~0.984) [43], PVdF-HFP/LiTFSI/EMIMFSI electrolyte (~0.99) [50], and PVA/LiTFSI/EMITFSI electrolyte (~0.995) [37]. To demonstrate the practicability of the polymer electrolyte, a closed-loop circuit composed of a LED bulb with the ILPE-Zn-4 electrolyte membrane was designed. The LED bulb was successfully lit and emitted a brilliant blue light (see Figure 4d). This phenomenon further indicates that the obtained ILPE-Zn-4 polymer electrolyte is an ionic conductor.

The electrochemical stability of the ILPE membrane is essential for its practical applications. Figure 5 displays the CV curves of the ILPE-Zn-4 electrolyte membranes at a scan rate of 10 mV s^{-1}. One can see that the current is stable within the potential range of −2.08 to 2.06 V (ESW: ~4.14 V), which signifies that the electrolyte membrane can work stably within this potential range. The current rising beyond this range may relate to the degeneration of the electrolyte membrane, manifested by the occurrence of an electrochemical reaction. Nevertheless, this ESW (~4.14 V) is acceptable for electrochemical device applications. In addition, there are a pair of small current humps at around ±1.3 V, which probably arise from the formation of ion pairs or other by-products during this process [51,52].

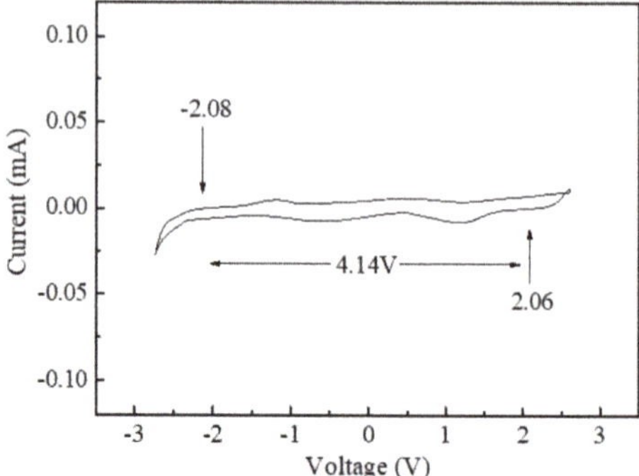

Figure 5. CV curve of the ILPE-Zn-4 membranes at a scan rate of 10 mV s^{-1}.

Thermal stability is a significant parameter for a polymer electrolyte membrane. A low thermal decomposing temperature in the polymer electrolyte may give rise to electrolyte membrane degradation in a short circuit, which may produce substantial heat accumulation. Hence, the electrolyte membranes should possess an excellent thermal stability. Figure 6 illustrates the TGA curves for ILPE membranes. It may be clearly observed that all polymer electrolytes show a slight drop between 80 to 180 °C. The weight losses of the membranes are: SPE-Zn ~4.07%, ILPE-Zn-1 ~3.93%, ILPE-Zn-2 ~3.80%, ILPE-Zn-3 ~4.03%, and ILPE-Zn-4 ~3.55%, which may be attributed to the evaporation of water absorbed in the ILPE membranes. However, a considerable decomposition of the EMITf-free SPE-Zn membrane may be observed when the temperature exceeds 350 °C. With the addition of ionic liquid, the polymer electrolyte membrane's decomposition temperature is lowered. This is primarily because of the enhancement of the polymer chain flexibility caused by the interaction of ionic liquid and Zn salt with the host polymer [53].

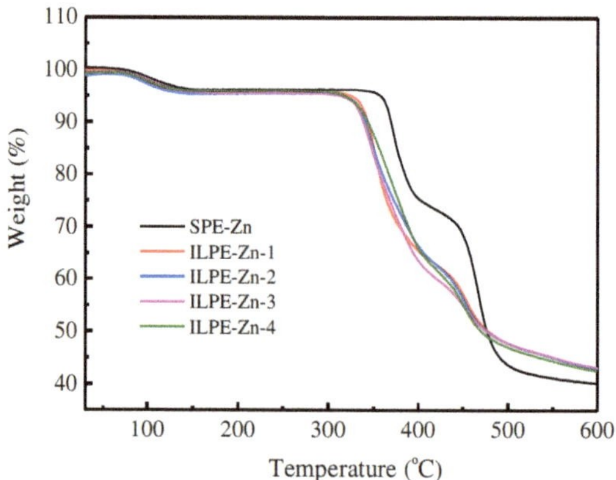

Figure 6. TGA curves of the ILPE membranes with different EMITf contents.

Despite this, the ILPE membranes incorporated with ionic liquid still display a sufficiently high thermal stability, and the decomposition temperature is approximately 305 °C. Therefore, in terms of the thermal stability, the polymer electrolyte membranes can be considered suitable for energy device applications.

In addition to a good thermal stability, the mechanical performance is also an imperative characteristic of electrolyte membranes. The typical stress-strain curves of SPE-Zn and PE-Zn-4 electrolyte membranes are shown in Figure 7. The values of the Young's modulus, tensile strength, and breaking strain are listed in the inset table. The Young's modulus and tensile strength of SPE-Zn are 220 MPa and 7.7 MPa, respectively, while the breaking strain is 380%. By contrast, those of ILPE-Zn-4 fall down to 117 MPa, 5.7 MPa, and 200%, respectively.

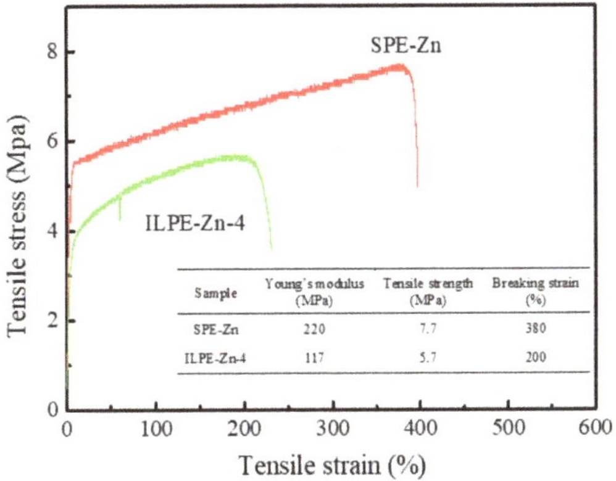

Figure 7. Typical stress-strain curves of SPE-Zn and ILPE-Zn-4 membranes and their mechanical properties (Young's modulus, tensile strength, and breaking strain).

The mechanical performance declines with the addition of EMITf. This may be due to a more amorphous phase and porous structure, created by the addition of EMITf in the ILPE-Zn-4 electrolyte membranes. In general, the polymer chains are highly flexible in the amorphous domains, thus reducing the interaction force between polymer molecules. Nevertheless, the electrolyte membranes still exhibit sufficient mechanical properties. This phenomenon has also been confirmed by other studies. As reported by Tang et al. [34], the electrolyte membrane (90PVdF-HFP:10Mg(Tf)$_2$ + 40EMITf) fabricated by solution casting possesses the following mechanical properties: Young's modulus = ~66 MPa, tensile strength = ~3.4 MPa, and breaking strain = ~633%. A study by Jie et al. [54] indicates that the tensile strength and breaking strain are about 2.2 MPa and 250%, respectively, for the PVDF-HFP/LiTFSI/NMP gel electrolyte film. Obviously, our values are highly comparable to theirs. In addition, Can et al. [55] developed a solid polymer electrolyte based on TPU/PEO = 1:3, showing that the electrolyte had a superior comprehensive performance with a tensile strength of 1.38 MPa and that it was successfully applied to the $LiFePO_4$/SPE/Li battery in the temperature range of 60 to 80 °C. Therefore, the mechanical performance of the ILPE-Zn-4 membrane should be good enough to be eligible for applications in energy storage devices.

4. Conclusions

Flexible ionic liquid-incorporated Zn-ion conducting polymer electrolyte membranes were prepared and characterized. Investigations indicate that the addition of ionic liquid EMITf reduces the crystallinity, enriches the nanopores' structure, and enhances the electrical and electrochemical properties of the electrolyte membranes. With a high thermal stability (thermal decomposition temperature ~305 °C) and good mechanical performance (tensile strength ~5.7 MPa), the optimized polymer electrolyte ILPE-Zn-4 (EMITf: Zn(Tf)$_2$: PVdF-HFP = 0.4: 0.4: 1 in mass) exhibits a high ionic conductivity (~1.44×10^{-4} S cm^{-1}) at an ambient temperature, with a wide electrochemical stability window (~4.14 V). Moreover, the ionic conductive polymer electrolyte exhibits an ultrahigh ion transfer number ~0.999. Therefore, the polymer electrolyte can be used as an ionic conductor to connect the circuit. The properties of the ionic conductor polymer electrolytes demonstrate that the optimized ionic liquid-incorporated Zn-ion conducting polymer electrolyte shows a promising perspective for energy storage applications.

Author Contributions: Conceptualization, J.L. and S.S.; Formal Analysis, J.L. and S.S.; Investigation, J.L. and S.A.; Data Curation, J.L. and Z.K.; Methodology, J.L. and T.W.; Writing—Original Draft Preparation, J.L.; Writing–Review & Editing, J.L., S.A., Z.K. and S.S. All authors have read and agreed to the published version of the manuscript.

Funding: This research was funded by the Open Project Program of Guangdong Provincial Key Laboratory of Electronic Functional Materials and Devices, Huizhou University (No. EFMD2020003Z).

Acknowledgments: The authors gratefully acknowledge the financial support for this study from the Open Project Program of Guangdong Provincial Key Laboratory of Electronic Functional Materials and Devices, Huizhou University (No. EFMD2020003Z).

Conflicts of Interest: The authors declare no conflict of interest.

References

1. Etacheri, V.; Marom, R.; Elazari, R.; Salitra, G.; Aurbach, D. Challenges in the development of advanced Li-ion batteries: A review. *Energy Environ. Sci.* **2011**, *4*, 3243–3262. [CrossRef]
2. Helbig, C.; Bradshaw, A.M.; Wietschel, L.; Thorenz, A.; Tuma, A. Supply risks associated with lithium-ion battery materials. *J. Clean. Prod.* **2018**, *172*, 274–286. [CrossRef]
3. Nitta, N.; Wu, F.; Lee, J.T.; Yushin, G. Li-ion battery materials: Present and future. *Mater. Today* **2015**, *18*, 252–264. [CrossRef]
4. Verma, V.; Kumar, S.; Manalastas, W.; Satish, R.; Srinivasan, M. Progress in Rechargeable Aqueous Zinc- and Aluminum-Ion Battery Electrodes: Challenges and Outlook. *Adv. Sustain. Syst.* **2019**, *3*, 1800111. [CrossRef]
5. Song, M.; Tan, H.; Chao, D.L.; Fan, H.J. Recent advances in Zn-ion batteries. *Adv. Funct. Mater.* **2018**, *28*, 1802564. [CrossRef]
6. Lee, B.; Yoon, C.S.; Lee, H.R.; Chung, K.Y.; Cho, B.W.; Oh, S.H. Electrochemically-induced reversible transition from the tunneled to layered polymorphs of manganese dioxide. *Sci. Rep.* **2015**, *4*, 6066. [CrossRef] [PubMed]
7. Yan, M.; He, P.; Chen, Y.; Wang, S.; Wei, Q.; Zhao, K.; Xu, X.; An, Q.; Shuang, Y.; Shao, Y.; et al. Water-Lubricated Intercalation in V_2O_5 center dot $nH_{(2)}O$ for High-Capacity and High-Rate Aqueous Rechargeable Zinc Batteries. *Adv. Mater.* **2018**, *30*, 1703725. [CrossRef] [PubMed]
8. Trocoli, R.; Mantia, F.L. An aqueous zinc-ion battery based on copper hexacyanoferrate. *ChemSusChem* **2015**, *8*, 481–485. [CrossRef] [PubMed]
9. He, P.; Yan, M.; Zhang, G.; Sun, R.; Chen, L.; An, Q.; Mai, L. Layered VS_2 Nanosheet-Based Aqueous Zn Ion Battery Cathode. *Adv. Energy Mater.* **2017**, *7*, 1601920. [CrossRef]
10. Kaveevivitchai, W.; Manthiram, A. High-capacity Zinc-ion Storage in An Open-Tunnel Oxide for Aqueous and Nonaqueous Zn-ion Batteries. *J. Mater. Chem. A* **2016**, *4*, 18737–18741. [CrossRef]
11. Zhang, N.; Cheng, F.; Liu, Y.; Zhao, Q.; Lei, K.; Chen, C.; Liu, X.; Chen, J. Cation-Deficient Spinel $ZnMn_2O_4$ Cathode in $Zn(CF_3SO_3)_2$ Electrolyte for Rechargeable Aqueous Zn-Ion Battery. *J. Am. Chem. Soc.* **2016**, *138*, 12894–12901. [CrossRef] [PubMed]
12. Sun, C.W.; Liu, J.; Gong, Y.D.; Wilkinson, D.P.; Zhang, J.J. Recent advances in all-solid-state rechargeable lithium batteries. *Nano Energy* **2017**, *33*, 363–386. [CrossRef]
13. Cheng, X.L.; Pan, J.; Zhao, Y.; Liao, M.; Peng, H.S. Gel Polymer Electrolytes for Electrochemical Energy Storage. *Adv. Energy Mater.* **2018**, *8*, 1702184. [CrossRef]
14. Zhu, M.; Wu, J.X.; Wang, Y.; Song, M.M.; Long, L.; Siyal, S.H.; Yang, X.P.; Sui, G. Recent advances in gel polymer electrolyte for high-performance lithium batteries. *J. Energy Chem.* **2019**, *37*, 126–142. [CrossRef]
15. Mindemark, J.; Lacey, M.J.; Bowden, T.; Brandell, D. Beyond PEO-alternative host materials for Li^+-conducting solid polymer electrolytes. *Prog. Polym. Sci.* **2018**, *81*, 114–143. [CrossRef]
16. Huang, Y.X.; Huang, Y.; Liu, B.; Cao, H.J.; Zhao, L.; Song, A.; Lin, Y.H.; Wang, M.S.; Li, X.; Zhang, Z.P. Gel polymer electrolyte based on p(acrylonitrile-maleic anhydride) for lithium ion battery. *Electrochim. Acta* **2018**, *286*, 242–251. [CrossRef]
17. Singh, S.K.; Shalu, B.L.; Gupta, H.; Singh, V.K.; Tripathi, A.K.; Verma, Y.L.; Singh, R.K. Improved electrochemical performance of EMIMFSI ionic liquid based gel polymer electrolyte with temperature for rechargeable lithium battery. *Energy* **2018**, *150*, 890–900. [CrossRef]
18. Song, J.Y.; Wang, Y.Y.; Wan, C.C. Conductivity study of porous plasticized polymer electrolytes based on poly(vinylidene fluoride)-A comparison with polypropylene separators. *J. Electrochem. Soc.* **2000**, *147*, 3219–3225. [CrossRef]

19. Li, M.S.; Liao, Y.H.; Liu, Q.Y.; Xu, J.X.; Sun, P.; Shi, H.N.; Li, W.S. Application of the imidazolium ionic liquid based nano-particle decorated gel polymer electrolyte for high safety lithium ion battery. *Electrochim. Acta* **2018**, *284*, 188–201. [CrossRef]
20. Francis, C.F.J.; Kyratzis, I.L.; Best, A.S. Lithium-Ion Battery Separators for Ionic-Liquid Electrolytes: A Review. *Adv. Mater.* **2020**, *32*, 1904205. [CrossRef]
21. Stephan, A.M. Review on gel polymer electrolytes for lithium batteries. *Eur. Polym. J.* **2006**, *42*, 21–42. [CrossRef]
22. He, C.F.; Liu, J.Q.; Cui, J.Q.; Li, J.; Wu, X.F. A gel polymer electrolyte based on Polyacrylonitrile/organic montmorillonite membrane exhibiting dense structure for lithium ion battery. *Solid State Ion.* **2018**, *315*, 102–110. [CrossRef]
23. Guo, Q.P.; Han, Y.; Wang, H.; Xiong, S.Z.; Liu, S.K.; Zheng, C.M.; Xie, K. Preparation and characterization of nanocomposite ionic liquid-based gel polymer electrolyte for safe applications in solid-state lithium battery. *Solid State Ion.* **2018**, *321*, 48–54. [CrossRef]
24. Saroj, A.L.; Singh, R.K. Thermal, dielectric and conductivity studies on PVA/Ionic liquid [EMIM][EtSO4] based polymer electrolytes. *J. Phys. Chem. Solids* **2012**, *73*, 162–168. [CrossRef]
25. Mohamed, N.S.; Arof, A.K. Investigation of electrical and electrochemical properties of PVDF-based polymer electrolytes. *J. Power Sources* **2004**, *132*, 229–234. [CrossRef]
26. Mccrum, N.G. A Study of Internal Friction in Copolymers of Tetrafluoroethylene and Hexafluoropropylene. *Makromolekul. Chem.* **1959**, *34*, 50–66. [CrossRef]
27. Abbent, S.; Plestil, J.; Hlavata, D.; Lindgren, J.; Tegenfeldt, J.; Wendsjo, A. Crystallinity and morphology of PVdF-HFP-based gel electrolytes. *Polymer* **2001**, *42*, 1407–1416. [CrossRef]
28. Long, L.Z.; Wang, S.J.; Xiao, M.; Meng, Y.Z. Polymer electrolytes for lithium polymer batteries. *J. Mater. Chem. A* **2016**, *4*, 10038–10069. [CrossRef]
29. Das, S.; Kumar, P.; Dutta, K.; Kundu, P.P. Partial Sulfonation of PVdF-co-HFP: A Preliminary Study and Characterization for Application in Direct Methanol Fuel Cell. *Appl. Energy* **2014**, *113*, 169–177. [CrossRef]
30. Hariprasad, R.; Vinothkannan, M.; Kim, A.R.; Yoo, D.J. SPVdF-HFP/SGO nanohybrid proton exchange membrane for the applications of direct methanol fuel cells. *J. Dispers. Sci. Technol.* **2019**, 1–13. [CrossRef]
31. Gnana, K.G.; Kim, A.R.; Nahm, K.S.; Yoo, D.J. High proton conductivity and low fuel crossover of polyvinylidene fluoride–hexafluoro propylene–silica sulfuric acid composite membranes for direct methanol fuel cells. *Curr. Appl. Phys.* **2011**, *11*, 896–902. [CrossRef]
32. Kim, A.R.; Gabunada, J.C.; Yoo, D.J. Sulfonated fluorinated block copolymer containing naphthalene unit/sulfonated polyvinylidene-co-hexafluoropropylene/functionalized silicon dioxide ternary composite membrane for low-humidity fuel cell applications. *Colloid Polym. Sci.* **2018**, *296*, 1891–1903. [CrossRef]
33. Kumar, D.; Kanchan, D.K. Dielectric and electrochemical studies on carbonate free Na-ion conducting electrolytes for sodium-sulfur batteries. *J. Energy Storage* **2019**, *22*, 44–49. [CrossRef]
34. Tang, X.; Muchakayala, R.; Song, S.H.; Zhang, Z.Y.; Polu, A.R. A study of structural; electrical and electrochemical properties of PVdF-HFP gel polymer electrolyte films for magnesium ion battery applications. *J. Ind. Eng. Chem.* **2016**, *37*, 67–74. [CrossRef]
35. Liu, J.H.; Khanam, Z.; Muchakayala, R.; Song, S.H. Fabrication and characterization of Zn-ion-conducting solid polymer electrolyte films based on PVdF-HFP/$Zn(Tf)_2$ complex system. *J. Mater. Sci. Mater. Electron.* **2020**, *31*, 6160–6173. [CrossRef]
36. Rathika, R.; Austin, S.S. Effect of ionic liquid 1-ethyl-3-methylimidazolium hydrogen sulfate on zinc-ion dynamics in PEO/PVdF blend gel polymer electrolytes. *Ionics* **2018**, *25*, 1137–1146. [CrossRef]
37. Wang, J.W.; Zhao, Z.J.; Song, S.H.; Ma, Q.; Liu, R.C. High Performance Poly(vinyl alcohol)-Based Li-Ion Conducting Gel Polymer Electrolyte Films for Electric Double-Layer Capacitors. *Polymers* **2018**, *10*, 1179. [CrossRef]
38. Liew, C.W.; Arifin, K.H.; Kawamura, J.; Iwai, Y.; Ramesh, S.; Arof, A.K. Effect of halide anions in ionic liquid added poly(vinyl alcohol)-based ion conductors for electrical double layer capacitors. *J. Non-Cryst. Solids* **2017**, *458*, 97–106. [CrossRef]
39. Senthil, K.P.; Sakunthala, A.; Reddy, M.V.; Prabu, M. Structural, morphological, electrical and electrochemical study on plasticized PVdF-HFP/PEMA blended polymer electrolyte for lithium polymer battery application. *Solid State Ion.* **2018**, *319*, 256–265. [CrossRef]

40. Tsao, C.H.; Kuo, P.L. Poly(dimethylsiloxane) hybrid gel polymer electrolytes of a porous structure for lithium ion battery. *J. Membr. Sci.* **2015**, *489*, 36–42. [CrossRef]
41. Yang, L.; Hu, J.Y.; Lei, G.; Liu, H.T. Ionic liquid-gelled polyvinylidene fluoride/polyvinyl acetate polymer electrolyte for solid supercapacitor. *Chem. Eng. J.* **2014**, *258*, 320–326. [CrossRef]
42. Rani, M.S.A.; Mohamed, N.S.; Isa, M.I.N. Investigation of the Ionic Conduction Mechanism in Carboxymethyl Cellulose/Chitosan Biopolymer Blend Electrolyte Impregnated with Ammonium Nitrate. *Int. J. Polym. Anal. Charact.* **2015**, *20*, 491–503. [CrossRef]
43. Tang, J.W.; Muchakayala, R.; Song, S.H.; Wang, M.; Kumar, K.N. Effect of $EMIMBF_4$ ionic liquid addition on the structure and ionic conductivity of $LiBF_4$-complexed PVdF-HFP polymer electrolyte films. *Polym. Test.* **2016**, *50*, 247–254. [CrossRef]
44. Monisha, S.; Selvasekarapandian, S.; Mathavan, T.; Milton, F.B.A.; Manoharan, S.; Karthikeyan, S. Preparation and characterization of biopolymer electrolyte based on cellulose acetate for potential applications in energy storage devices. *J. Mater. Sci. Mater. Electron.* **2016**, *27*, 9314–9324. [CrossRef]
45. Samsudin, A.S.; Saadiah, M.A. Ionic conduction study of enhanced amorphous solid bio-polymer electrolytes based carboxymethyl cellulose doped NH_4Br. *J. Non-Cryst. Solids* **2018**, *497*, 19–29. [CrossRef]
46. Unnisa, C.N.; Chitra, S.; Selvasekarapandian, S.; Monisha, S.; Devi, G.N.; Moniha, V.; Hema, M. Development of poly(glycerol suberate) polyester (PGS)–PVA blend polymer electrolytes with NH4SCN and its application. *Ionics* **2018**, *24*, 1979–1993. [CrossRef]
47. Ravi, M.; Song, S.H.; Wang, J.W.; Wang, T.; Nadimicherla, R. Ionic liquid incorporated biodegradable gel polymer electrolyte for lithium ion battery applications. *J. Mater. Sci. Mater. Electron.* **2016**, *27*, 1370–1377. [CrossRef]
48. Wang, J.W.; Song, S.H.; Muchakayala, R.; Hu, X.C.; Liu, R.C. Structural, electrical, and electrochemical properties of PVA-based biodegradable gel polymer electrolyte membranes for Mg-ion battery applications. *Ionics* **2017**, *23*, 1759–1769. [CrossRef]
49. Hu, X.C.; Muchakayala, R.; Song, S.H.; Wang, J.W.; Chen, J.J.; Tan, M.L. Synthesis and optimization of new polymeric ionic liquid poly(diallydimethylammonium) bis(trifluoromethane sulfonyl)imde based gel electrolyte films. *Int. J. Hydrogen Energy* **2018**, *43*, 3741–3749. [CrossRef]
50. Singh, S.K.; Gupta, H.; Balo, L.; Singh, V.K.; Tripathi, A.K.; Verma, Y.L.; Singh, R.K. Electrochemical characterization of ionic liquid based gel polymer electrolyte for lithium battery application. *Ionics* **2018**, *24*, 1895–1906. [CrossRef]
51. Vraneš, M.; Cvjetićanin, N.; Papović, S.; Šarac, B.; Prislan, I.; Megušar, P.; Gadžurić, S.; Rogač, B.M. Electrical, electrochemical and thermal properties of the ionic liquid + lactone binary mixtures as the potential electrolytes for lithium-ion batteries. *J. Mol. Liq.* **2017**, *243*, 52–60. [CrossRef]
52. Wang, J.W.; Chen, G.P.; Song, S.H. Na-ion conducting gel polymer membrane for flexible supercapacitor application. *Electrochim. Acta* **2020**, *330*, 135322. [CrossRef]
53. Pal, P.; Ghosh, A. Highly efficient gel polymer electrolytes for all solid-state electrochemical charge storage devices. *Electrochim. Acta* **2018**, *278*, 137–148. [CrossRef]
54. Jie, J.; Liu, Y.L.; Cong, L.N.; Zhang, B.H.; Lu, W.; Zhang, X.M.; Liu, J.; Xie, H.M.; Sun, L.Q. High-performance PVDF-HFP based gel polymer electrolyte with a safe solvent in Li metal polymer battery. *J. Energy Chem.* **2020**, *49*, 80–88. [CrossRef]
55. Tao, C.; Gao, M.H.; Yin, B.H.; Li, B.; Huang, Y.P.; Xu, G.; Bao, J.J. A promising TPU/PEO blend polymer electrolyte for all-solid-state lithium ion batteries. *Electrochim. Acta* **2017**, *257*, 31–39. [CrossRef]

Article

From Cellulose, Shrimp and Crab Shells to Energy Storage EDLC Cells: The Study of Structural and Electrochemical Properties of Proton Conducting Chitosan-Based Biopolymer Blend Electrolytes

Shujahadeen B. Aziz [1,2,*], Muhamad. H. Hamsan [3], Muaffaq M. Nofal [4], Saro San [5], Rebar T. Abdulwahid [1,6], Salah Raza Saeed [7], Mohamad A. Brza [1,8], Mohd F. Z. Kadir [9], Sewara J. Mohammed [10] and Shakhawan Al-Zangana [11]

1 Hameed Majid Advanced Polymeric Materials Research Lab., Department of Physics, College of Science, University of Sulaimani, Qlyasan Street, Sulaimani 46001, Kurdistan Regional Government, Iraq; rebar.abdulwahid@univsul.edu.iq (R.T.A.); mohamad.brza@gmail.com (M.A.B.)
2 Department of Civil Engineering, College of Engineering, Komar University of Science and Technology, Sulaimani 46001, Kurdistan Regional Government, Iraq
3 Institute for Advanced Studies, University of Malaya, Kuala Lumpur 50603, Malaysia; hafizhamsan93@gmail.com
4 Department of Mathematics and General Sciences, Prince Sultan University, P. O. Box 66833, Riyadh 11586, Saudi Arabia; muaffaqnofal@gmail.com
5 Department of Physics and Astronomy, University of Missouri-Kansas City, MO 64110, USA; ssawcc@mail.umkc.edu
6 Department of Physics, College of Education, University of Sulaimani, Old Campus, Sulaimani 46001, Kurdistan Regional Government, Iraq
7 Charmo Research Center, Charmo University, Peshawa Street, Chamchamal, Sulaimani 46001, Kurdistan Regional Government, Iraq; salah.saeed@charmouniversity.org
8 Manufacturing and Materials Engineering Department, Faculty of Engineering, International Islamic University of Malaysia, Kuala Lumpur 50603, Gombak, Malaysia
9 Centre for Foundation Studies in Science, University of Malaya, Kuala Lumpur 50603, Malaysia; mfzkadir@um.edu.my
10 Department of Chemistry, College of Science, University of Sulaimani, Qlyasan Street, Sulaimani 46001, Kurdistan Regional Government, Iraq; sewara.mohammed@univsul.edu.iq
11 Department of Physics, College of Education, University of Garmyan, Kalar 46021, Kurdistan Regional Government, Iraq; shakhawan.al-zangana@garmian.edu.krd
* Correspondence: shujahadeenaziz@gmail.com

Received: 17 June 2020; Accepted: 6 July 2020; Published: 9 July 2020

Abstract: In this study, solid polymer blend electrolytes (SPBEs) based on chitosan (CS) and methylcellulose (MC) incorporated with different concentrations of ammonium fluoride (NH_4F) salt were synthesized using a solution cast technique. Both Fourier transformation infrared spectroscopy (FTIR) and X-ray diffraction (XRD) results confirmed a strong interaction and dispersion of the amorphous region within the CS:MC system in the presence of NH_4F. To gain better insights into the electrical properties of the samples, the results of electrochemical impedance spectroscopy (EIS) were analyzed by electrical equivalent circuit (EEC) modeling. The highest conductivity of 2.96×10^{-3} S cm^{-1} was recorded for the sample incorporated with 40 wt.% of NH_4F. Through transference number measurement (TNM) analysis, the fraction of ions was specified. The electrochemical stability of the electrolyte sample was found to be up to 2.3 V via the linear sweep voltammetry (LSV) study. The value of specific capacitance was determined to be around 58.3 F/g. The stability test showed that the electrical double layer capacitor (EDLC) system can be recharged and discharged for up to 100 cycles with an average specific capacitance of 64.1 F/g. The synthesized EDLC cell was found to exhibit high efficiency (90%). In the 1st cycle, the values of internal resistance, energy density and power density of the EDLC cell were determined to be 65 Ω, 9.3 Wh/kg and 1282 W/kg, respectively.

Keywords: polymer blend; XRD and FTIR; impedance study; TNM and LSV; CV and EDLC

---

1. Introduction

Energy storage devices, such as lithium batteries, supercapacitors and fuel cells using liquid electrolytes, have attracted significant attention in recent years, owing to their ionic nature. However, there are several issues that still need to be solved, such as the release of harmful gases, the lack of safety and corrosive action [1]. It is difficult to use a harmless liquid in energy storage devices without any drawbacks. Designing a desirable device with a proper size and shape that fits liquid electrolytes is challenging [2]. This encourages scientists and researchers to work on the development of a safe and efficient solid polymer electrolyte (SPE). SPEs can provide satisfactory thermal stability, low weight, high flexibility, cost effectiveness and easy handling [3]. On the other hand, the harmful effects of plastic wastes on the environment are recognized to cause global warming and water pollution. Therefore, there is a special interest in the development of biodegradable and biocompatible natural polymers as SPEs [4]. This is due to their abundance, biocompatibility, biodegradability and cost effectiveness [5]. These favorable properties have made scientific circles extensively utilize natural polymers in polymer electrolyte-based devices [6]. There are many natural polymers, including starch, CS, carboxymethyl cellulose (CMC), MC and rubber that can be used in the synthesis of SPEs [5,6]. In this study, CS and MC were used as natural polymers. CS is often extracted from crustaceans (crabs, lobsters, crayfish, shrimp, krill and barnacles) and has a chemical structure of β-(1→4)2-amino-2-deoxy-D-glucose-(D-glucosamine) [4], while MC is obtained from mixing alkali-based cellulose with methyl chloride. The chemical structure of MC comprises a 1,4 glycosidic bond [7]. The reduced ion mobility in SPE matrices has led numerous research groups to develop different approaches that have improved the ambient conductivity. Two common approaches which are widely addressed are the blending of two polymers and using a variety of salts [8–13]. Polymer blending is regarded as a promising technique to upgrade the properties of individual polymer constituents. Many new and enhanced characteristics can be achieved through the polymer blending technique, such as relatively high ionic conductivity, flexibility, transference number and thermal stability [14]. Recently, polymer electrolytes assembled from biopolymer attracted the attention of many research groups due to their availability for a wide range of applications in electrochemical devices [2–4,8,10–12]. Both CS and MC are known to contain functional groups with lone pair electrons that assist ion transport within their matrixes. This is due to the fact that the lone pair electrons within their structure can serve as complexation sites for the ions.

The preparation of proton (H^+)-conducting SPEs is usually involves mixing strong inorganic acids or ammonium salts. For instance, sulfuric acid (H_2SO_4) and phosphoric acid (H_3PO_4) are the two commonly used inorganic acids. However, the main drawback of these inorganic acids is their chemical degradation when mixed with SPEs, which leads to incompatibility with practical applications [15]. Therefore, the ammonium salts are usually utilized to obtain a proton-conducting SPE with a relatively high ionic conductivity and thermal stability. The continuous interactions of the charge carriers with the available functional group then generates motion of the polymer chain segments and thus makes the polymer more conductive [16]. Radha et al. have documented a conductivity value of 6.9×10^{-6} S/cm for the polyvinyl alcohol (PVA)–ammonium fluoride (NH_4F) system [17]. Additionally, the enhancement in the dielectric behavior of PVA with the presence of NH_4F was also addressed. Another research study has also recently reported a relatively high conductivity of 6.40×10^{-7} S/cm for the MC–NH_4F system at room temperature [5]. Doping NH_4F into the CS–dextran system also resulted in a high value of conductivity, up to 10^{-3} S/cm [18].

Electrical double layer capacitors (EDLCs) are one of the promising electrochemical energy storage devices that fulfill the requirements of high power application with fast charge–discharge cycles. The energy storage mechanism in these devices involves charge accumulation on the carbon electrode surface at the interfacial region in the form of potential energy [19]. EDLCs feature a fantastic high power density, long cycle life, fast charge–discharge rate and simplistic fabrication procedure [20]. In EDLC devices, various types of carbon have been used as electrode materials, for instance, graphite [21], aerogel [22], carbon nanofibers [23] and activated carbon [24]. The most used one is activated carbon, which is almost an ideal active material that is defined by a high surface area, satisfactory electronic conductivity and cost effectiveness [25]. Based on an earlier study, ammonium salts have been shown to exhibit reasonable proton donor behavior if incorporated into polymer matrices [26]. In order to enhance the performance of biodegradable-based EDLC devices to meet the industrial level, various polymer blended electrolytes with different dopant salts were studied. An extensive literature survey revealed that the effect of NH_4F salt concentration on the conductivity of a CS:MC blended system and its use in EDLC devices had not already been investigated. Thus, for this work, firstly, systems of CS:MC incorporated with various concentrations of NH_4F were examined through electrical and structural analyses. Then, the relatively highest conducting SPBE film was utilized as the electrode separator in an EDLC device application.

2. Experimental Part

2.1. Materials and Sample Preparation

In this work, CS with a relatively high molecular mass of around 310,000 to 375,000 g/mol was used, along with MC and NH_4F in the fabricating of SPBE systems. All the materials were supplied by Sigma-Aldrich Corporation (Missouri, MO, USA). Firstly, for the preparation of CS:MC polymer blend electrolytes, two separated solutions of CS and MC with percentages of 70 wt.% and 30 wt.%, respectively, were dissolved in 40 mL of 1% acetic acid. They were stirred for 3 hrs at room temperature. Based on previous work [27], this ratio of CS and MC was shown to be optimal in the preparation of CS:MC polymer blend electrolytes. Then, both solutions of CS and MC were mixed with continuous stirring for 2 hrs in order to obtain a final homogeneous blended solution. Subsequently, with continuous stirring, different portions of NH_4F, ranging from 10 to 40 wt.% in steps of 10 wt.%, were added separately to a series of blended solutions to obtain CS:MC:NH_4F electrolytes. The final solutions were then poured into Petri dishes to cast films at ambient temperature. For further drying, the formed films were then transferred into a desiccator to achieve solvent-free films. The obtained polymer blend electrolyte samples were coded as CMCF0, CMCF1, CMCF2, CMCF3 and CMCF4 for CS:MC doped with 0, 10, 20, 30 and 40 wt.% of NH_4F, respectively.

2.2. Structural and Impedance Analyses

XRD measurements were performed to study the structural properties of the samples, using a Siemens D5000 X-ray diffractometer (1.5406 Å) (Bruker AXS GmbH, Berlin, Germany). The acquisition process comprised scanning the 2θ angle continuously from 5° to 80° (resolution = 0.1°). FTIR (FT-IR, Spotlight 400 Perkin-Elmer spectrometer, Waltham, MA, USA) was conducted in the range 450 to 4000 cm^{-1} with a 1 cm^{-1} resolution. Electrical properties of the samples were studied by means of EIS (HIOKI 3532–50 LCR Hi-TESTER) (Hioki, Nagano, Japan) in the frequency range of 50 Hz to 5 MHz. For the purpose of electrical characterizations, the samples were sandwiched between two stainless steel electrodes. Based on the results obtained in EIS measurements and the sample dimensions, the conductivities of the samples were determined, using the following equation:

$$\sigma_{dc} = \left(\frac{1}{R_b}\right) \times \left(\frac{t}{A}\right) \tag{1}$$

where R_b is the bulk resistance of the sample, t is the film thickness and A is the surface area of the film.

2.3. Transference Number Measurement (TNM) and Linear Sweep Voltammetry (LSV) Studies

To use polymer blend electrolytes in applications, it is important to study their TNM and LSV measurements. Through TNM measurements, one can verify the ion dominancy in the conduction process. LSV measurements were used to investigate the sample electrolytes' potential stability. The TNM measurements were performed by using a V & A Instrument DP3003 digital DC power supply with 0.20 V (V & A Instrument, Shanghai, China). LSV measurements were also carried out by using a Digi-IVY DY2300 Potentiostat at a sweep rate of 50 mV s^{-1} (Neware, Shenzhen, China). The cell used in both TNM and LSV measurements was composed of two stainless steel (SS) disks and the highest conducting SPBE film. The cell was then mounted in a Teflon holder, as shown in Figure 1.

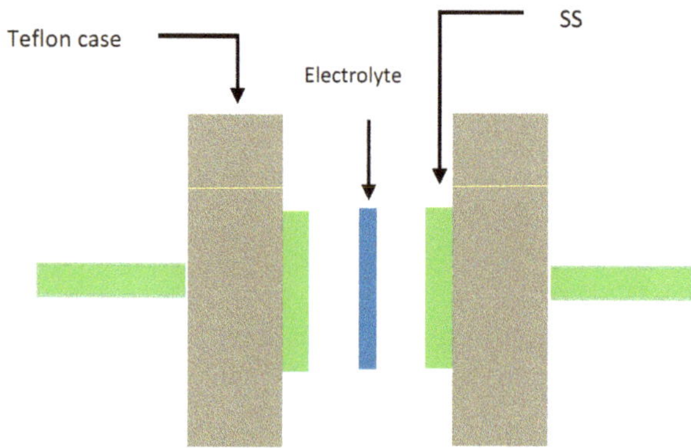

Figure 1. Schematic appearance of the cell used for the LSV and TNM measurements.

2.4. EDLC Preparation

For the fabrication of an EDLC device, electrodes composed of polyvinylidene fluoride (PVdF), activated carbon and carbon black materials, were used. Under medium stirring, 0.5 g of PVdF was dissolved in 15 mL of N-methyl pyrrolidone (NMP) to obtain an NMP–PVdF solution. The activated carbon and carbon black materials were dry mixed for 15 min by using a planetary ball miller (XQM-0.4) with a rotational speed of 500 r/min. The obtained powders were composed of 3.25 g and 0.25 g of activated carbon and carbon black materials, respectively. The powder was then poured into the NMP–PVdF solution and stirred to dissolve completely. Then, the mixture was cast on an aluminum foil with a doctor blade technique to obtain a thick black solution. Subsequently, the folded solution was heated at 60 °C in an oven for a certain time to obtain a dried state. The obtained dried bulk electrode was then cut into a circle with an area of 2.01 cm^2. In the final step of the EDLC cell preparation, the relatively high conducting electrode film was sandwiched between two carbon electrodes and packed in CR2032 coin cells. The fabricated EDLC cell was then fixed over the Teflon holder for further testing. Cyclic voltammetry of the EDLC was performed at 10 mV s^{-1} and charged up to 0.90 V, by using a Digi-IVY DY2300 Potentiostat (Neware, Shenzhen, China). Furthermore, the charge–discharge profiles of the EDLC were also examined using a Neware battery cycler (Neware, Shenzhen, China) with a current density of 0.2 mA cm^{-2}.

3. Result and Discussion

3.1. XRD and FTIR Study

Figure 2 shows the XRD pattern of the pure and doped CM:MC samples. Previous studies have illustrated that pure MC displays broad humps and several weak peaks at $2\theta = 8°$ and 21° [28]. It has also been revealed that CS in its pure state, which has a predominant crystalline phase, is characterized by two obvious peaks at 14.50° and 20.90° as a result of inter- and intra-hydrogen bonding [29]. It is obvious in Figure 2a that the XRD peaks due to pure CS almost disappeared and only a hump can be seen. This indicates that the technique of polymer blending is a novel approach to overcome the crystalline phases. This is due to the formation of hydrogen bonding between MC and CS matrices. It is well defined that hydrogen bonding is based on the interaction between electron-deficient hydrogen and a high electron density region. In fact, hydrogen bonding (H-bonding), as an intermolecular interaction, can be expressed as X–H . . . Y, where X and Y are electronegative elements and Y possesses one or more lone electron pairs; in other words, X and Y are F, O and N atoms, respectively [30]. It is evident from the molecular structure of MC (see Figure 3) that the monomer of MC contains O atoms, which enables it to build H-bonding. Clearly, hydrogen bonding as a secondary force is much weaker than the primary bond within the molecules, such as covalent bonds and other polar bonds, but far stronger than the van der Waals interaction [30]. It is notable from Figure 3a that about six hydrogen bonds can be formed between CS monomers; meanwhile, only three intermolecular hydrogen bonds can be developed between the CH and MC monomer, as depicted in Figure 3b. Therefore, the blended samples showed a smaller number of hydrogen bonding sites, which resulted in the reduction of the degree of crystallinity. The XRD peak broadness of the CS:MC blend sample pinpoints that the inter-chain spacing in the blended sample became larger than that of the individual polymers. The expanded inter-chain spacing in the blend simplifies the dipole reorientation to the applied field due to high amorphous content. The addition of NH_4F to the blend electrolytes (BEs) caused a clear reduction in the intensity of XRD peaks. Interestingly, upon the addition of 40 wt.% of NH_4F, the lowest peak intensity was recorded. It is evidenced that the broad peak at 20.25° (see Figure 2c) emphasizes the amorphous nature of the system. The characteristic feature of the amorphous nature of a polymer body is a broad peak in the form of a hump. Increasing salt in such a polymer system caused a relative reduction in the intensity of the broad peak between 11° and 27.21°, thus increasing the amorphous structure of the BE system [26]. Therefore, the XRD pattern can certainly be used to show the amorphous characteristic of the samples, which was reflected in the gradual decrease in intensity with peak broadening due to the addition of NH_4F salt. It is also distinct that the dissimilarity in the intensity and sharpness of the XRD peaks of the polymer electrolytes after salt addition can be a good confirmation of strong interactions between the polymer and the inorganic salt [29].

FTIR is considered as an effective technique to deal with a new compound, in terms of both structure and composition, that forms during a chemical reaction. The extent of interaction between CS and MC in the BPE system was confirmed via the FTIR technique. Figure 4a–c shows the FTIR spectra of CS:MC biopolymer electrolytes in the wave number range of 4000–890 cm^{-1}. It is motivating to observe both position shifting and intensity variation of the bands, which are considered as evidence of the existence of particular functional groups in pure CS:MC and CS:MC:NH_4F electrolyte systems. A more important observation is the confirmation of the presence of heteroatoms (e.g., O and N) with lone pair electrons in a desired fabricated polymer host as electrolytes [31]. The appearance of a strong peak at around 2900 cm^{-1} is ascribed to the C-H stretching modes, as shown in Figure 4c [32,33], and its intensity decreased with rising salt concentration. It is also seen from the same figure that the CS polymer is characterized by a single -NH_2 group and a couple of -OH groups in the repeating unit [34]. The doping process substantially affected the -OH stretching broad peak at 3359 cm^{-1} in Figure 4c [32]. The strong interaction between dopant salt and the CS:MC host blended polymer can be identified from both peak shifting and intensity changes. Based on earlier work [34], the presence of vibrational frequency peaks of -NH_2, O=C-NHR and –OH are considered as the characteristic FTIR spectra of MC and CS polymers. It is clear from Figure 4b that the shifting occurred towards lower wave numbers in the bands of amino NH_2, O=C-NHR and OH groups, confirming a strong interaction between the NH_4F dopant salt and the CS:MC host blended polymer.

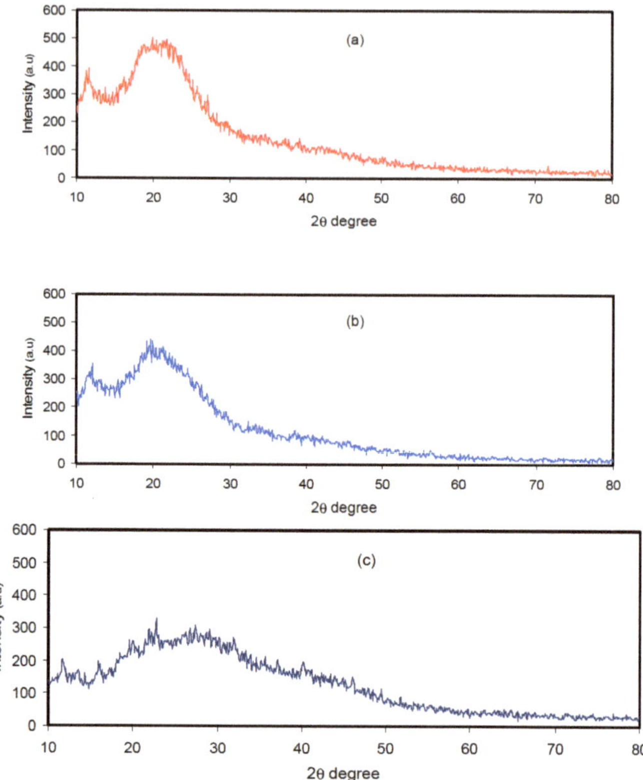

Figure 2. XRD pattern for (**a**) pure CS:MC, (**b**) CMCF2 and (**c**) CMCF4 blend electrolytes.

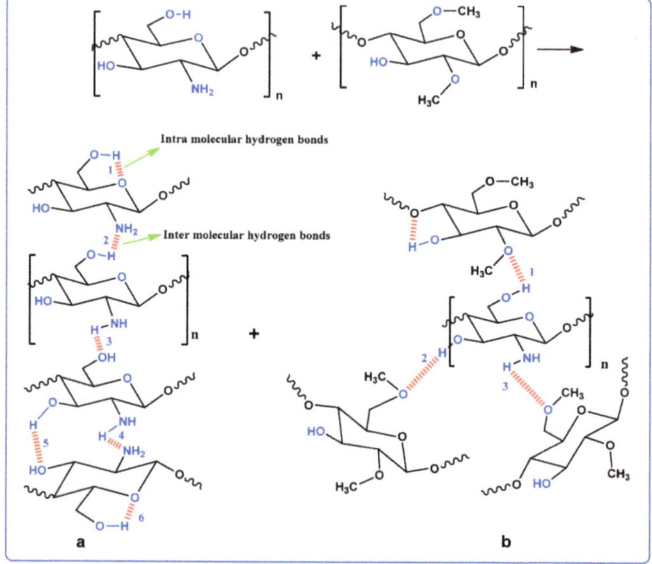

Figure 3. CS and MC polymer hydrogen bonding presentation; (**a**) hydrogen bonding through CS polymer and (**b**) hydrogen bonding CS:MC polymer blends.

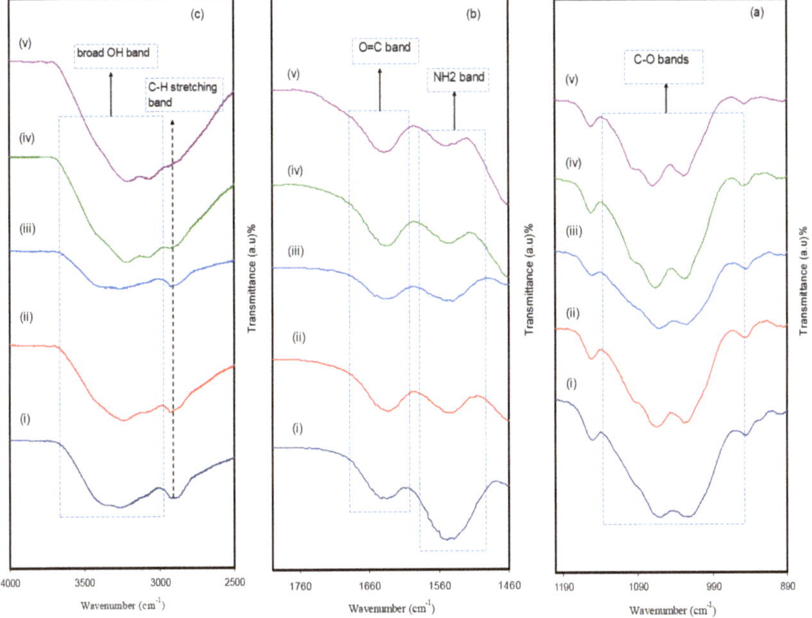

Figure 4. FTIR spectrum of (**i**) CMCF0, (**ii**) CMCF1, (**iii**) CMCF2, (**iv**) CMCF3 and (**v**) CMCF4 in the range (**a**) 890 cm^{-1} to 1190 cm^{-1}, (**b**) 1460 cm^{-1} to 1760 cm^{-1} and (**c**) 2500 cm^{-1} to 4000 cm^{-1}.

In Figure 4a, a peak at 1055 cm^{-1} appeared as a result of the antisymmetric stretching of an asymmetric oxygen bridge in the cyclohexane ring and a range from 1150 to 1000 cm^{-1} is attributed to the C-O-C bond [28]. It is observable that the intensity of the peaks decreased and to some extent shifting occurred as a consequence of the increment of NH_4F concentration. These certify that there is a strong interaction between the polymer body and amine salt via coordination bonds and thus confirms the complexation [35]. The addition of NH_4F salt provides cations that attract oxygen atoms at the C-O-C ether group in the identical polymer to produce polymer salt complexes [28]. In the current polymer salt system, the $NH^+{}_4$ ion from NH_4F coordinates to both the O atom of the ether group and the hydroxyl group in the CS and MC host polymer blend. These interactions prove the occurrence of protonation in the present electrolytes. It is clearly verified in the shifting of hydroxyl, ether, C=O and -NH_2 [5].

3.2. Impedance Study

The impedance spectra of the blend electrolyte films at ambient temperature are shown in Figure 5a–d. The semicircle at the high frequency region can be related to the parallel combination of the bulk electrolyte resistance (R_E) and the bulk electrolyte capacitance (C_E), owing to the migration process of proton ions and the immobilized state of polymer chains, respectively. Interestingly, the semicircle diameter was lessened with increasing salt concentration. This implies that the relaxation of ions occurred at different times [36]. The electrode/electrolyte capacitance (C_{EE}) produced by the accumulated double-layer ions at the electrode/electrolyte interface (i.e., the low frequency spike region) is represented by another capacitor in series with the parallel combination of a resistor and capacitor corresponding to the high frequency semicircle. The schematic illustration of EEC for Figure 5a,b is shown in Figure 6. To confirm our interpretation, the experimental impedance data were simulated with EECs, as can be seen in Figure 7. It is worth noting that at a high frequency region, the semicircle completely disappeared for 30 and 40 wt.% of NH_4F salt concentrations, as clearly shown in Figure 5c,d.

This suggested that only the resistive component of the obtained impedance spectra is responded to by the polymer host body [27]. In this case, the resistor of the blend electrolyte film in series with the capacitor of the double-layer capacitances represents the electrical behavior of the system. The conductivity improvement upon the addition of 40 wt.% of NH_4F can be attributed to the increase in the number of mobile charge carriers. Moreover, the amorphous nature of the polymer electrolyte could have a vital role. Consequently, it results in an inferior energy barrier and facilitates the ion transport [35]. It is self-evident that the ionic conductivity of an electrolyte depends on both the number and mobility of ions, as can be seen from the following equation [37]:

$$\sigma = \sum \eta q \mu \tag{2}$$

where the carrier density is denoted as η, elementary charge is symbolized by q and μ is the mobility. The literature confirmed that in a polymer–ammonium salt system, the charge-carrying species is an H^+ ion that is offered by an ammonium ion [27]. The most general theory of proton conduction is structure diffusion, which is known as the Grotthuss mechanism, where ion exchanging occurs between the complexed sites [38]. Proton conduction by the Grotthus mechanism states that protons jump over the complexing sites, leading to the creation of a vacant site followed by reorientation to occupy the vacant site [27]. Equation (1) was used to calculate the DC conductivity of the pure CS:MC and CS:MC:NH_4F electrolyte samples at room temperature. Table 1 lists the DC conductivities of the samples. It is noteworthy that the DC conductivity increased from 7.16×10^{-10} S cm^{-1} for pure CS:MC to 7.34×10^{-4} S cm^{-1} for CS:MC incorporated with 40 wt.% of NH_4F. Previous studies have confirmed that polymer electrolytes with high DC conductivity, ranging from 10^{-5} to 10^{-3} S cm^{-1}, can be crucial for electrochemical device applications, including batteries and EDLCs.

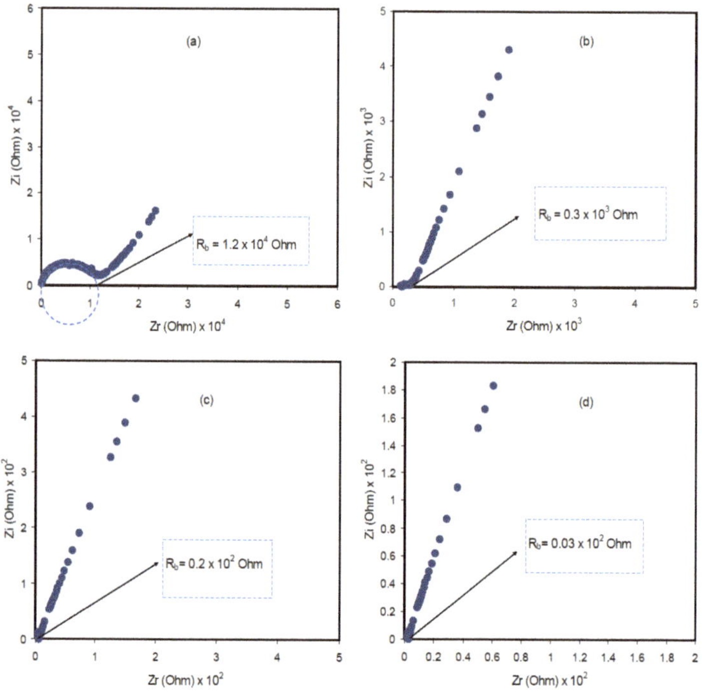

Figure 5. Impedance plots of (**a**) CMCF1, (**b**) CMCF2, (**c**) CMCF3 and (**d**) CMCF4 blended films at ambient temperature.

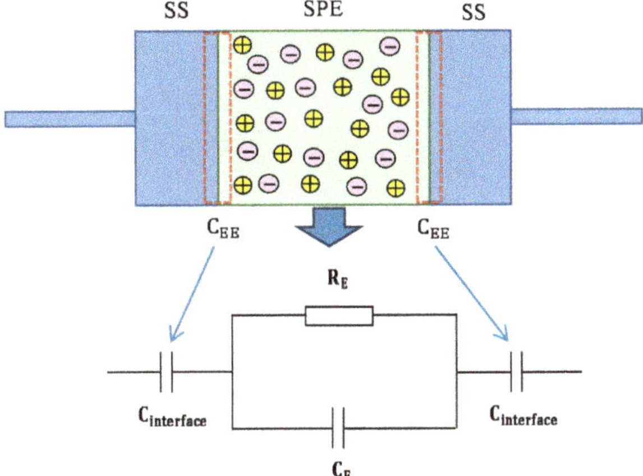

Figure 6. Schematic illustration of electrical equivalent circuits (EECs) for impedance plots consisting of a high-frequency semicircle and low-frequency tails.

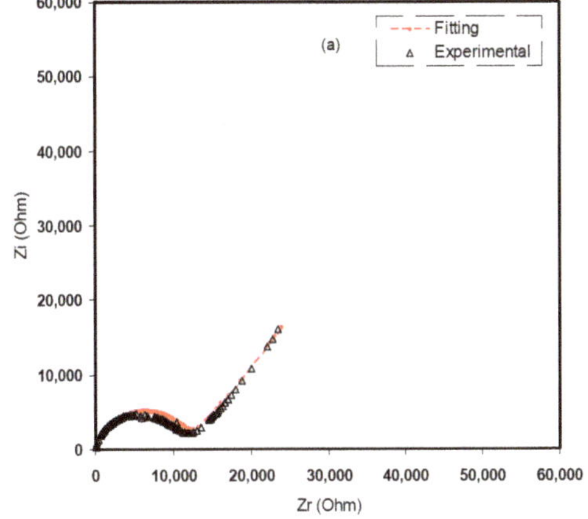

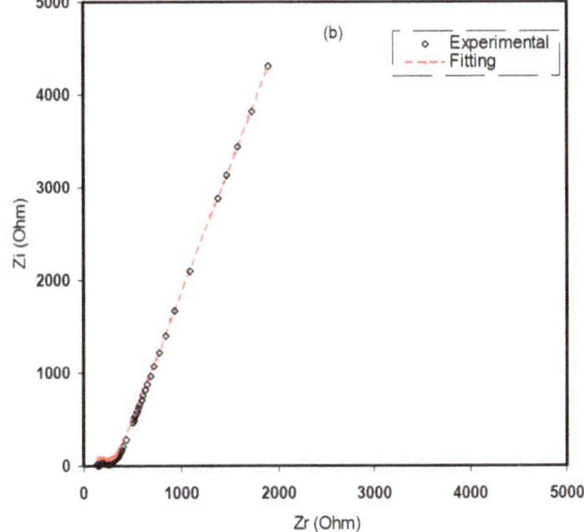

Figure 7. *Cont.*

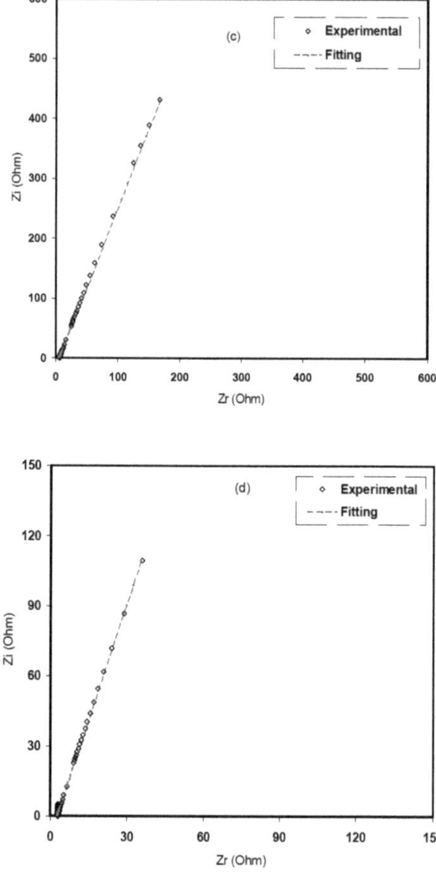

Figure 7. Experimental impedance and EEC fitting plots for (**a**) CMCF1, (**b**) CMCF2, (**c**) CMCF3 and (**d**) CMCF4 blended films at ambient temperature.

Table 1. DC conductivity for pure CS:MC and blend electrolyte films at room temperature.

Sample Designation	DC Conductivity (S cm^{-1})
CMCF1	8.37×10^{-10}
CMCF1	7.42×10^{-7}
CMCF2	2.96×10^{-5}
CMCF3	4.45×10^{-4}
CMCF4	2.96×10^{-3}

From the fitting and analysis of experimental spectra using the EEC technique, one can easily deal with the mechanism of the system under study [39]. The experimental impedance plots with EEC modeling for all the electrolyte samples are shown in Figure 7a–d. This model guides us to calculate and understand the electrical properties of solid-based electrolyte polymers. A three-component equivalent circuit reflects the experimental impedance plots. To be precise, the three principal elements are ZCPE1 (constant phase, electrode/electrolyte capacitance, $C_{interface}$), ZCPE2 (another constant phase, electrolyte capacitance, C_E) and a bulk resistance R_b for the SPE (bulk electrolyte resistance, R_E) (see Figure 8a). Two elements, R_b and ZCPE1, are obtained in the high frequency region; in other words, both respond in the high frequencies, whereas the low frequency spike region is linked to ZCPE2.

The EECs corresponding to Figure 7a,b are shown in Figure 8a. The ZCPE1 is the response of the double-layer capacitance formed at the interface region between the electrodes and the SPE [40]. It is possible to derive ZCPE's impedance as [41]:

$$Z_{CPE} = \frac{1}{C\omega^p}\left[\cos\left(\frac{\pi p}{2}\right) - i\sin\left(\frac{\pi p}{2}\right)\right] \tag{3}$$

where CPE is constant phase element, capacitance is represented by C, the angular frequency is denoted by ω and p is related to the departure of the plot from the vertical axis in complex impedance plots. It is worth mentioning that CPE is the acronym most commonly used instead of capacitor in the context of an EEC model. It is appealing to see that in such a system, the capacitor element in the circuit is replaced by CPE, where it is a capacitor system in an ideal or pure capacitor. Stated differently, this implies that the system is a semi-capacitor [40,42], which is the nature of a capacitor in the electrolyte/electrode system. This is not the case to recognize an ideal capacitor system in the existing experimental impedance spectra. For the case of expressing the real (Zr), imaginary (Zi) and complex impedance (Z*) values in the equivalent circuit, it can be formulated in a mathematical expression in the following way [41]:

$$Z_r = \frac{R_b C_1 \omega^{p1}\cos\left(\frac{\pi p_1}{2}\right) + R_b}{2R_b C_1 \omega^p \cos\left(\frac{\pi p}{2}\right) + {R_b}^2 C^2 \omega^{2p} + 1} + \frac{\cos\left(\frac{\pi p_2}{2}\right)}{C_2 \omega^{p2}} \tag{4}$$

$$Z_i = \frac{R_b C_1 \omega^{p1}\sin\left(\frac{\pi p_1}{2}\right)}{2R_b C_1 \omega^p \cos\left(\frac{\pi p}{2}\right) + {R_b}^2 C^2 \omega^{2p} + 1} + \frac{\sin\left(\frac{\pi p_2}{2}\right)}{C_2 \omega^{p2}} \tag{5}$$

where R_b is the bulk resistance. Based on Equations (4) and (5), the experimental impedance plots are well simulated, as shown in Figure 7, and the EECs are presented in Figure 8. From the impedance spectra, it is obviously seen that the semicircle size dropped at the high frequency region as the concentration of salt increased (20 wt.% of NH_4F). Figure 7a,b exhibits a model comprising the incomplete semicircle, from which one can extract the value of R_b that is parallel with the CPE element and series with another CPE relating to a low frequency tail, as shown schematically in Figure 8a. Predictably, the incomplete semicircle at 30 wt.% and 40 wt.% of NH_4F totally disappeared, as shown in Figure 7c,d. This indicates the possibility of the resistive behavior of SPEs and the CPE component in series, as shown schematically in Figure 8b. The equivalent circuit element parameters of the blend electrolytes are shown in Table 2. The logical explanation for this result, where the semicircle disappeared at the high frequency region in the spectra, is explained by the entire conductivity attributed to a huge ion migration [43]. In this case, the values of Z_r and Z_i are correlated to the EEC and can be expressed mathematically as follows:

$$Z_r = \frac{\cos\left(\frac{\pi p}{2}\right)}{C\omega^p} + R_b \tag{6}$$

$$Z_i = \frac{\sin\left(\frac{\pi p}{2}\right)}{C\omega^p} \tag{7}$$

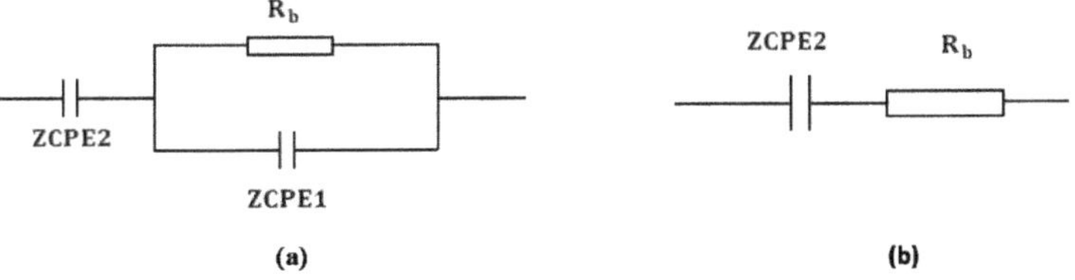

Figure 8. Schematic illustration of the EECs for (**a**) parallel combination of a resistor and capacitor in series with another capacitor and (**b**) series combination of a resistor and capacitor. The resistor is represented by the symbol —▭— and the capacitor is represented by —||—.

Table 2. The parameters of the circuit elements of the blend electrolytes.

Electrolytes	p_1 (rad)	p_2 (rad)	k_1 (F^{-1})	k_2 (F^{-1})	C_1 (F)	C_2 (F)
CMCF1	0.89	0.60	7.00×10^9	2.90×10^6	1.43×10^{-10}	3.45×10^{-7}
CMCF2	0.56	0.77	4.00×10^6	2.55×10^6	2.50×10^{-7}	3.92×10^{-7}
CMCF3	-	0.77	-	2.10×10^5	-	4.76×10^{-6}
CMCF4	-	0.81	-	1.13×10^5	-	8.85×10^{-6}

k is the inverse of C ($k = 1/C$).

3.3. EDLC Characteristics

TNM and LSV Study

To study total conductivity in SPBEs, TNM analysis was determined and the dominancy of ions to the total conductivity was verified. It was also proven that electrons partly contributed to the overall conductivity. Therefore, both electronic (t_e) and ionic (t_i) transference numbers can be obtained via:

$$t_i = \frac{I_i - I_s}{I_i} \quad (8)$$

$$t_i = 1 - t_e \quad (9)$$

where I_i is the current at the initial stage and I_s is the current at the constant stage. Figure 9 displays the TNM plot for the highest conducting SPE. From the procedure, as the cell was disturbed by the working voltage of 0.2 V, the value of I_i was obtained at 15.7 A. The high value of the current at the initial stage was assigned to both ions and electrons as charge-carrying species. As the procedure continued, a dramatic drop in the current was seen before reaching a constant value of 4.2 A. Obviously, this decrease in the initial total current is ascribed to the depletion of the ionic species in the bulk electrolyte and became constant in the completely depleted state [43]. This phenomenon is explained on the basis of the ion-blocking effect at the stainless-steel electrodes. Kufian et al. [44] clarified this behavior and stated that, as polarization occurs in the cell, a constant current stage reaches its value and the cause of the remaining current flow is related to electrons alone. The values of t_e and t_i were determined to be 0.27 and 0.73, respectively. This finding is of major importance, since it clarifies the main contribution of ions compared to electrons in the total conductivity. These data results are in good harmony with those reported for the carboxylmethylcellulose–NH_4F system by Ramlli and Isa [45]. Kyle et al. [46] recommended that the high value of the ionic transference number verifies to a large extent that the SPE behaves as an ionic conductor.

Among a number of characteristics of SPEs, electrochemical stability is critical, and from this one can decide the viability of SPEs in electrochemical devices. In this work, the potential window extended to ~1 V [47]. Figure 10 exhibits the LSV for the relatively highly conducting SPE. Electrolytes used as electrode separators in EDLC devices will be subjected to a continuous process of rapid charge–discharge. During the charging process, a high voltage will be produced and the electrolyte film will breakdown. Thus, it is crucial to charge the EDLC to a potential value well below the breakdown voltage. Within the potential range from 0 to 1.7 V, there was no significant current change as the potential swept. Despite the current rising beyond 1.7 V, it was not considerable. As the potential exceeded 2.3 V, the current rose significantly, indicating electrolyte decomposition at the surface of the inert electrodes [48]. Remarkably, it is realized that the CS:MC:NH_4F system is stable electrochemically within the specified potential range. Previously reported work [49] for an NH_4F-based CS:dextran system found an electrochemical stability up to 1.7 V. Therefore, the CS:MC:NH_4F system can be utilized in EDLC fabrication.

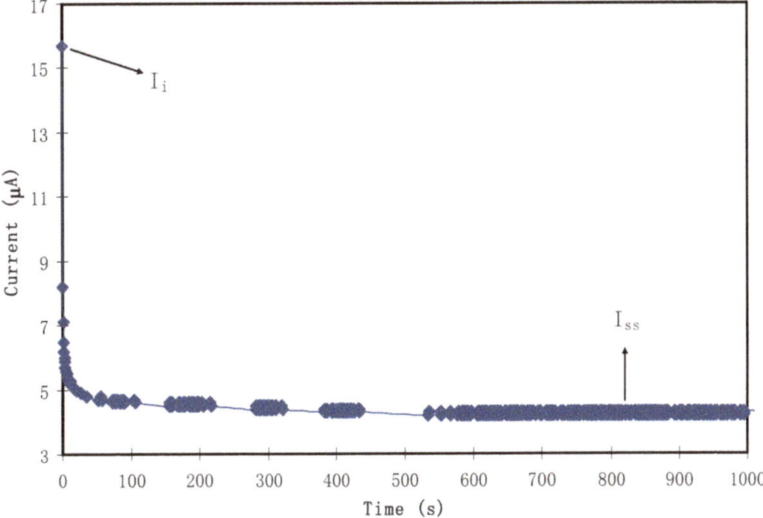

Figure 9. TNM plot for the highest conducting solid polymer electrolyte (SPE) with a polarization voltage of 0.2 V.

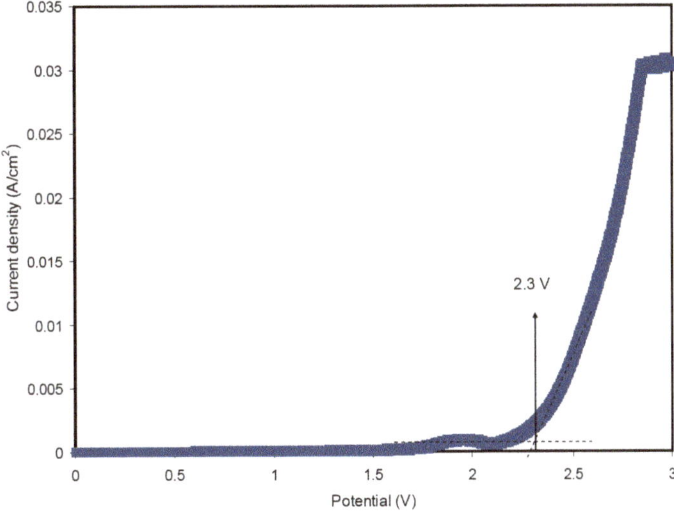

Figure 10. LSV plot for the highest conducting SPE with a scan rate of 50 mV/s.

3.4. EDLC

To investigate capacitive behavior of an EDLC, it is the preliminary test to record cyclic voltammetry (CV). Figure 11a highlights the CV of the assembled EDLC at a sweep rate of 10 mV/s. The schematic diagram for CV measurement is shown in Figure 11b and the realist image of the prepared electrolyte after the LSV test is presented in Figure 11c. It is noticed that the shape of the CV is almost rectangular with the absence of any redox peaks. This is a good sign of the pure EDLC (completely capacitor) and there was no signature for the existence of pseudocapacitors in the energy storage system [50]. Therefore, the non-Faradaic charge storage comprises ion adsorption and accumulation at the interfacial region rather than an intercalation/deintercalation process. More clearly, both electrons and ions accumulate at the interface of electrodes and electrolyte, respectively. The charge accumulation in this region makes a double-layer charge regime in the form of potential energy [51]. Typically, an ideal capacitor manifests in a perfect rectangular shape of CV. The non-ideal rectangular shape of our CV might be caused by both internal resistance and electrode porosity [52]. To analyze the obtained results, it is calculable to find the specific capacitance (C_{spe}) of the assembled EDLC from the CV via the following equation:

$$C_{spe} = \int_{V_i}^{V_f} \frac{I(V)dV}{2mv\left(V_f - V_i\right)} \quad (10)$$

where $I(V)dV$ is the area of the CV which is determined using Origin 9.0 software through the integration function. The chosen V_i and V_f in the present work are 0 V and 0.9 V, respectively, and m and v are the mass of used active material and sweep rate, respectively. The value of C_{spe} extracted from the CV was 58.3 F/g. This value is comparable to that obtained from the charge–discharge graph.

At a current density of 0.2 mA/cm^2, the rechargeability of the assembled EDLC is depicted in Figure 12a. From the figure, it can be seen that the plot is approximately linear in a triangular shape, which is a capacitive characteristic of the EDLC [53]. The results from the galvanostatic technique are in a high harmony with those from the capacitive CV analysis, confirming the capacitive behavior. It is straightforward to calculate specific capacitance (C_s) from the slope (s) of the charging and discharging curve using the following relationship:

$$C_s = \frac{i}{sm} \quad (11)$$

where i is the applied current, which was 0.4 mA in the present work.

Figure 12b presents C_s for the complete 100 cycles. C_s was 82.3 F/g for the 1st cycle. The relatively high value of C_s in the initial few cycles was caused by the rapid development of the potential double layer due to electron and ion accumulation at the electrode and the electrolyte interfacial region. The value dropped to 78.4 F/g in the 5th cycle. This decline in specific capacitance value with increasing cycle number might be caused by ion association to form ion pairs or ion aggregation, which blocked the migration of free ions towards the carbon electrode [54]. Then, the value of C_s stabilized from the 10th cycle to the 100th cycle with an average C_s value of 64.1 F/g. This value was nearly the same as that obtained from the CV analysis. The study of the EDLC of various active materials and their corresponding specific capacitances are presented in Table 3.

Another two important parameters are efficiency (η) and equivalent series resistance (ESR), which were determined for the assembled EDLC in 100 cycles. The η value can be computed easily from the discharging (t_{dis}) and charging (t_{cha}) times, as shown in Figure 13a using the following equation:

$$\eta = \frac{t_{dis}}{t_{cha}} \times 100 \quad (12)$$

From Figure 13a, one can see clearly that the η in the 1st cycle was 43% and increased remarkably to 75% in the 5th cycle. The lower value of recorded η in the initial cycle might owe to the longer duration of the charging process, where the ions and electrons built the charge double layer at the surface of the carbon electrodes/electrolyte. Almost the maximum value of 92% was reached in the 30th cycle, which then became constant at an average of 92.1% up to the 80th cycle. At the steady state, charging time was almost the same as the discharging time, which is ideal for a typical capacitor. Interestingly, the value of η decreased to 91.5% and 90.4% in the 90th and 100th cycles, respectively. The efficiency decline of the EDLC system resulted from the development of internal resistance. It is worth mentioning that η was observed to be harmonized with C_s, when it lowered from 65.8 F/g to 90.1 F/g in the 90th and 100th cycles, respectively. Shukur et al. [58] pointed out that a satisfactory EDLC must have 90–95% η. It was also claimed that a relatively high value of efficiency reflects a compatible electrolyte–electrode contact.

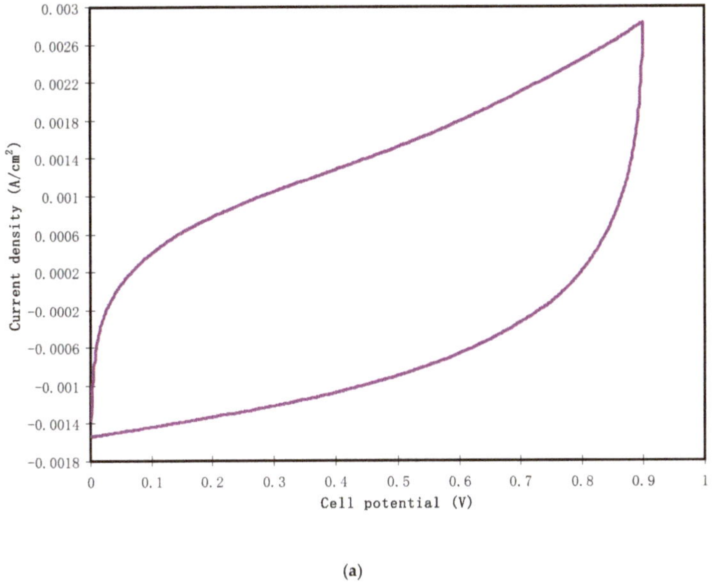

(a)

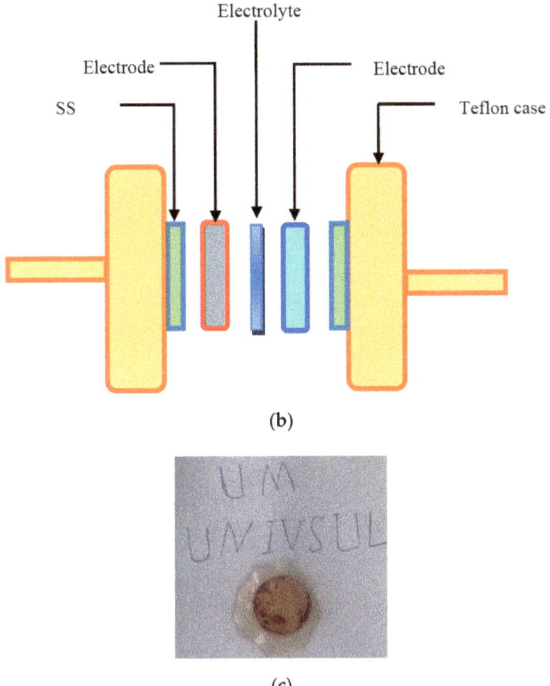

(b)

(c)

Figure 11. (**a**) CV measurement for the highest conducting SPE at 10 mV/s from 0 to 0.9 V, (**b**) schematic diagram of the CV measurement unit for the fabricated electric double-layer capacitor (EDLC) cell, and (**c**) realist image of the highest conducting SPE when reaching the breakdown voltage.

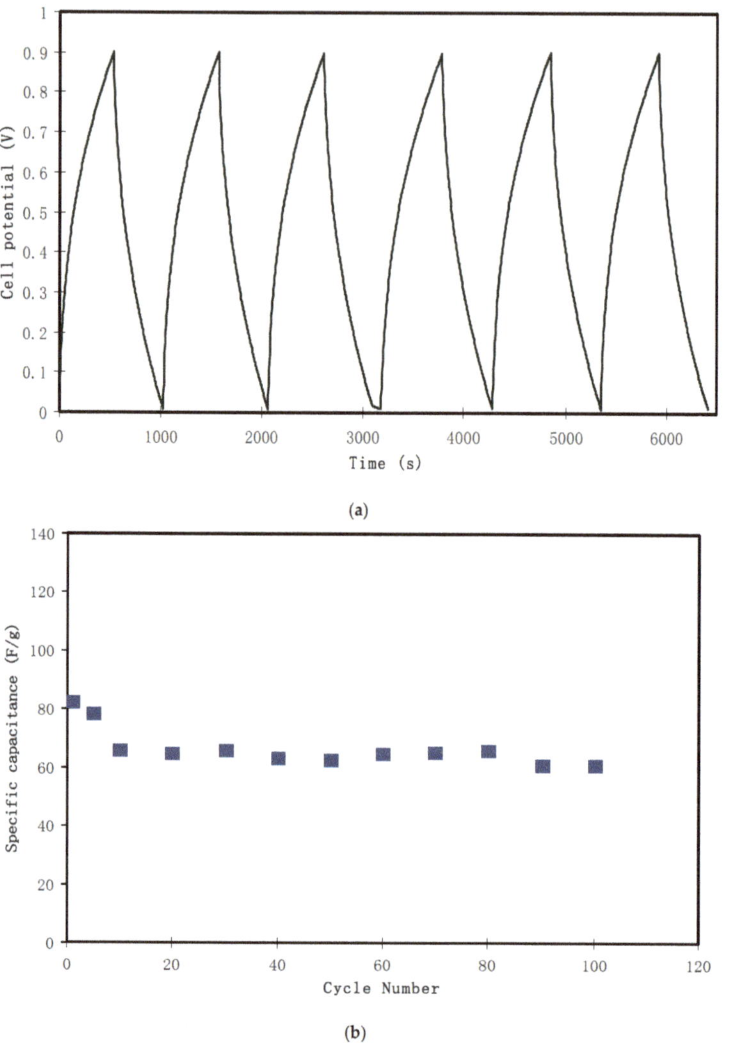

Figure 12. (**a**) Charge–discharge profile for the assembled EDLC at 0.2 mA/cm^2, (**b**) specific capacitance of the assembled EDLC for 100 cycles.

Table 3. EDLC studies with various active materials and their specific capacitances.

System	Active Materials	C_s (F/g)	Reference
CS:poly(ethylene oxide) (PEO):NH_4SCN	Activated carbon	3.8	[55]
Poly(vinyl alcohol)(PVA):dextran:NH_4I	Activated carbon	4.2	[56]
CS: MC:NH_4I	Activated carbon	6.9	[8]
CS:Dextran:NH_4I	Activated carbon	19.1	[10]
Hydroxylethyl cellulose + $MgTf_2$ + EMIMT + silica nanoparticles	Activated carbon	25.1	[1]
PVA + CH_3COONH_4 + BmImCl	Activated carbon	31.3	[3]
MC + NH_4NO_3 + PEG	PEG/Activated carbon	38	[50]
PVA/polystyrene	Carbon	40	[57]
Cellulose + Na_2SO_4	Cellulose nanofiber + graphite	43	[21]
CS/MC + NH_4F	**Activated carbon**	**64.1**	**This work**

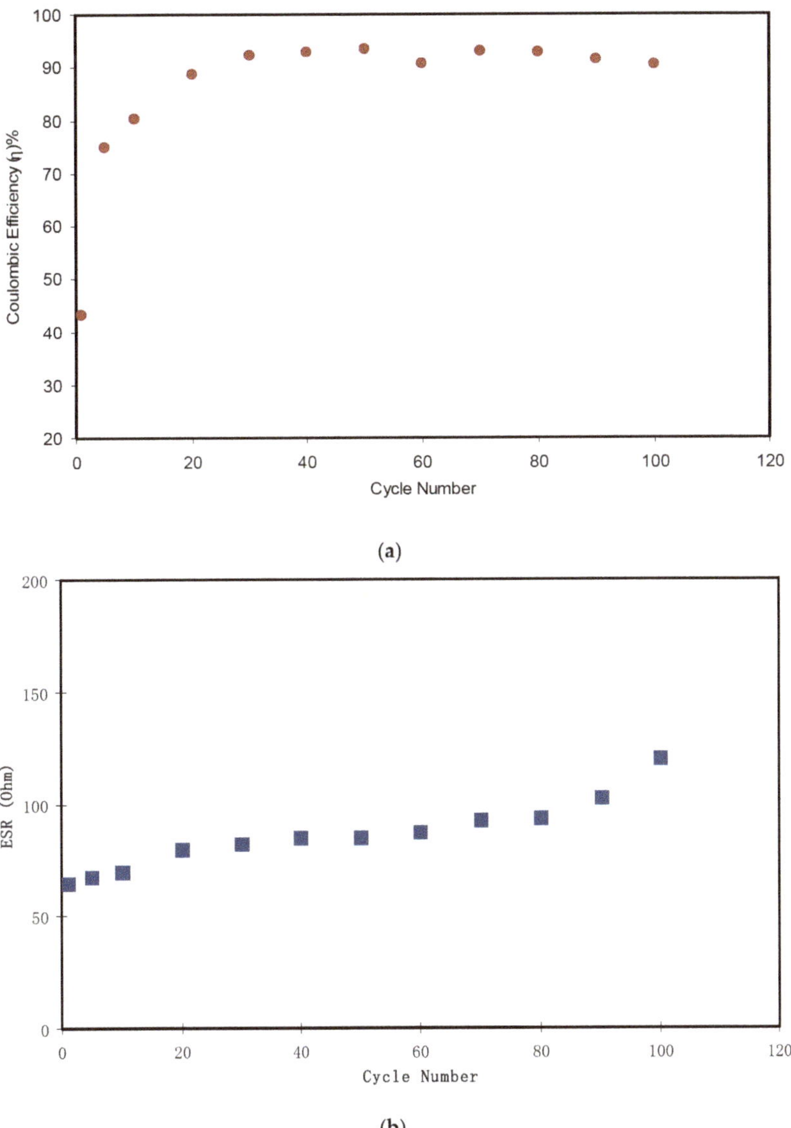

Figure 13. (**a**) Efficiency of the assembled EDLC for 100 cycles and (**b**) internal resistance of the assembled EDLC for 100 cycles.

Figure 12a showed that the drop voltage (V_{drop}) in the charge–discharge profile starts insignificantly prior to the discharging process starting in the assembled EDLC system. This V_{drop} of the assembled EDLC in the present study was quantized, ranging from 0.026 to 0.048 V. This drop in potential might be related to the development of internal resistance in the bulk electrolyte, which is called equivalent series resistance (*ESR*) and can be calculated from the following relationship:

$$ESR = \frac{V_{drop}}{i} \quad (13)$$

The equivalent series resistance (*ESR*) of the EDLC for the complete 100 cycles is exhibited in Figure 13b. From the figure, the value of *ESR* in the 1st cycle was 65 Ω and increased to 75 Ω in the 10th cycle. One of the interesting observations is an almost constant value of *ESR* from the 20th to the 80th cycle, with an average value of 86.6 Ω. Furthermore, the ESR value increased to 102 Ω and 120 Ω as the cycle number was increased to 90 and 100, respectively. A harmonized trend is seen in the pattern of C_s and η, where it is almost constant from the 20th cycle to the 80th cycle and starts to drop at the 90th cycle. Herein, three main factors are discussed that caused this increase in the internal resistance of the EDLC. Firstly, ion aggregation was formed from a rapid charge–discharge process. Secondly, the electrode–electrolyte gap resulted in an increase in the internal resistance. Thirdly, there was a technical issue caused by the fabrication of carbon electrodes on the aluminum current collector [59].

The crucial parameters in the investigation of an EDLC are energy (E_{den}) and power (P_{den}) density. The energy density (Wh/kg) shows how much energy can be stored by an EDLC, whereas power density (W/kg) is a measure of the energy or power that can be delivered by an EDLC [60]. Simply, both E_{den} and P_{den} can be obtained from the equations shown below:

$$E_{den} = \frac{C_s V}{2} \quad (14)$$

$$P_{des} = \frac{V^2}{4m(ESR)} \quad (15)$$

Figure 14a,b presents the energy density and power density of the fabricated EDLC for 100 cycles, respectively. The applied voltage (V) on the assembled EDLC in the present work was 0.9 V and the magnitudes of both E_{den} and P_{den} in the 1st cycle were 9.3 Wh/kg and 1282 W/kg, respectively. These magnitudes dropped to 7.3 Wh/kg and 1041 W/kg at the 20th cycle for the energy density and power density, respectively. From the data analysis, it was recorded that the energy density magnitude was almost constant at an average of 7.3 Wh/kg and the power density magnitude remained nearly constant at 964 W/kg from the 20th cycle to the 80th cycle. The key observation is that the magnitudes of energy and power density exhibited a significant lowering in the 90th and 100th cycles.

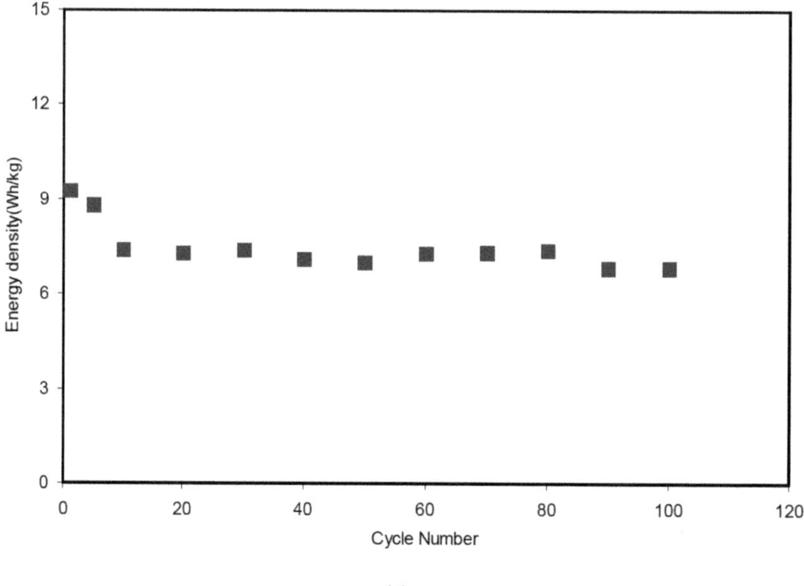

(a)

Figure 14. *Cont.*

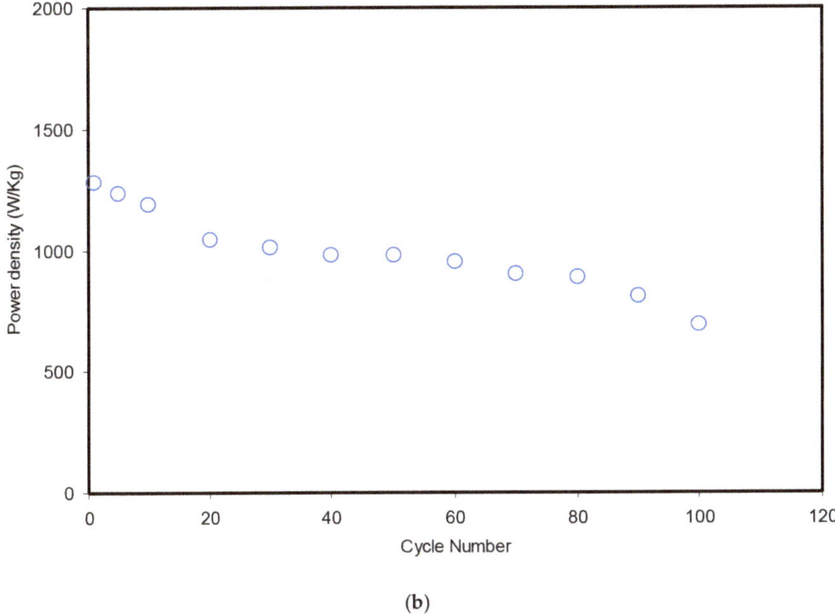

(b)

Figure 14. (**a**) Energy density of the assembled EDLC for 100 cycles and (**b**) power density of the assembled EDLC for 100 cycles.

These patterns harmonized with the patterns of C_s, η and ESR. From an earlier study [18], the EDLC of the CS:dextran:NH_4F electrolyte as the electrode separator reached magnitudes of 1.4 Wh/kg and 428 W/kg for E_{den} and P_{den}, respectively. The higher magnitudes of both the energy and power density recorded for the current work might be due to the lower extent of crystallinity of MC compared to dextran. In other words, the high ion mobility governs these crucial parameters. Furthermore, even from other studies, the degree of the crystallinity of dextran was reported to be 35.92 [61], while for MC, it was 32.89 [62]. It emphasized to a large extent that the conduction of ions is favorable in the amorphous region [63]. Ultimately, it is realized that the ion conduction in the CS:MC system is greater than in the CS:dextran host towards the electrode surface. Therefore, the more ions that diffuse and migrate towards the electrode surface, the larger the development of charge double layers, which in turn results in higher energy storage, and better performance of the EDLC. A consistent value of energy density during the ion transport towards the carbon electrodes confirms that ions in the CS:MC:NH_4F electrolyte experience the same energy barrier [64]. Generally, the EDLC performance decline based on C_s, η, E_{den} and P_{den} magnitudes is due to electrolyte depletion. It is worth mentioning that the rapid charge–discharge process causes free ions to recombine back to from aggregated ions, and consequently electrolyte depletion takes place. Eventually, the depletion of electrolyte phenomenon lowers the potential energy growth at the surface of the carbon electrodes [65–70].

4. Conclusions

In this work, a solid polymer blend electrolyte (SPBE), based on CS and MS incorporated with different amounts of NH_4F, was prepared. The possibility of employing the prepared SPBE in EDLC fabrication was investigated and the performance of the device was analyzed. The FTIR spectra revealed that there was a strong interaction between CS, MS and even NH_4F as a dopant. Both the peak shifting and intensity reduction of the XRD pattern were a good indication of the interaction between the components of the system. The structural analysis of the system confirmed the prevalence of an amorphous region. From EIS, the improvement of conductivity with increasing salt concentration was highlighted due to a rise in the number of mobile charge carriers. The charge carrying in total conduction mainly depended on ions, whereas electrons were a less common contributor, as t_e and t_i were determined to be 0.27 and 0.73, respectively. The rectangular shape of the CV revealed the presence of pure EDLC with capacitive behavior and recorded a high specific capacitance of 64.1 F/g. The high efficiency (above 90%) proved a good electrode–electrolyte contact. The suitability of the EDLC system was justified via energy and power density analysis. The relatively low value (86.6 Ω) of equivalent series resistance (ESR) over 80 cycles supported the good performance of the fabricated EDLC. Overall, the high conduction and electrochemical stability are behind the obtained relatively high energy and power density of 7.3 Wh/kg and 964 W/kg, respectively. Finally, it was concluded that the high conduction, electrochemical stability, low equivalent series resistance and amorphous nature played crucial roles in the good performance of the EDLC cell.

Author Contributions: Conceptualization, S.B.A., M.M.N., S.R.S., M.F.Z.K., S.J.M. and S.A.-Z.; Formal analysis, S.B.A. and M.H.H.; Investigation, S.B.A. and M.H.H.; Methodology, S.B.A. and M.H.H.; Project administration, S.B.A. and M.M.N.; Validation, S.S., R.T.A., M.A.B. and S.A.-Z.; Visualization, S.J.M.; Writing—original draft, S.B.A. and M.H.H.; Writing—review and editing, M.M.N., S.S., R.T.A., S.R.S., M.A.B., M.F.Z.K. and S.A.-Z. All authors have read and agreed to the published version of the manuscript.

Funding: The authors gratefully acknowledge the financial support for this study from the Ministry of Higher Education and Scientific Research-Kurdish National Research Council (KNRC), Kurdistan Regional Government/Iraq and the Prince Sultan University.

Acknowledgments: The authors gratefully acknowledge the financial support for this study from Ministry of Higher Education and Scientific Research, Kurdish National Research Council (KNRC), Kurdistan Regional Government/Iraq. The financial support from the University of Sulaimani and Komar Research Center (KRC), Komar University of Science and Technology are greatly appreciated.

Conflicts of Interest: The authors declare no conflict of interest.

References

1. Chong, M.Y.; Numan, A.; Liew, C.W.; Ng, H.; Ramesh, K.; Ramesh, S. Enhancing the performance of green solid-state electric double-layer capacitor incorporated with fumed silica nanoparticles. *J. Phys. Chem. Solids* **2018**, *117*, 194–203. [CrossRef]
2. Ramaswamy, M.; Malayandi, T.; Subramanian, S.; Srinivasalu, J.; Rangaswamy, M.; Soundararajan, V. Development and Study of Solid Polymer Electrolyte Based on Polyvinyl Alcohol: Mg (ClO 4) 2. *Polym. Technol. Eng.* **2017**, *56*, 992–1002. [CrossRef]
3. Liew, C.W.; Ramesh, S.; Arof, A.K. Good prospect of ionic liquid-based poly (vinyl alcohol) polymer electrolytes for supercapacitors with excellent electrical, electrochemical and thermal properties. *Int. J. Hydrogen Energy* **2014**, *39*, 2953–2963. [CrossRef]
4. Shujahadeen, B.A.; Brza, M.A.; Salah, R.S.; Hamsan, M.H.; Kadir, M.F.Z. Ion association as a main shortcoming in polymer blend electrolytes based on CS: PS incorporated with various amounts of ammonium tetrafluoroborate. *J. Mater. Res. Technol.* **2020**, *9*, 5410–5421. [CrossRef]
5. Aziz, N.A.N.; Idris, N.K.; Isa, M.I.N.M. Solid Polymer Electrolytes Based on Methylcellulose: FT-IR and Ionic Conductivity Studies. *Int. J. Polym. Anal. Charact.* **2010**, *15*, 319–327. [CrossRef]
6. Saadiah, M.; Zhang, D.; Nagao, Y.; Muzakir, S.; Samsudin, A. Reducing crystallinity on thin film-based CMC/PVA hybrid polymer for application as a host in polymer electrolytes. *J. Non-Cryst. Solids* **2019**, *511*, 201–211. [CrossRef]
7. Taghizadeh, M.T.; Seifi-Aghjekohal, P.; Taghizadeh, M.T. Sonocatalytic degradation of 2-hydroxyethyl cellulose in the presence of some nanoparticles. *Ultrason. Sonochem.* **2015**, *26*, 265–272. [CrossRef]
8. Aziz, S.B.; Hamsan, M.; Brza, M.; Kadir, M.; Muzakir, S.; Abdulwahid, R.T. Effect of glycerol on EDLC characteristics of chitosan: Methylcellulose polymer blend electrolytes. *J. Mater. Res. Technol.* **2020**, *9*, 8355–8366. [CrossRef]
9. Mathew, C.M.; Karthika, B.; Ulaganathan, M.; Rajendran, S. Electrochemical analysis on poly (ethyl methacrylate)-based electrolyte membranes. *Bull. Mater. Sci.* **2015**, *38*, 151–156. [CrossRef]

10. Aziz, S.B.; Hamsan, M.H.; Nofal, M.M.; Karim, W.O.; Brevik, I.; Brza, M.; Abdilwahid, R.T.; Al-Zangana, S.; Kadir, M.F.Z. Structural, Impedance and Electrochemical Characteristics of Electrical Double Layer Capacitor Devices Based on Chitosan: Dextran Biopolymer Blend Electrolytes. *Polymers* **2020**, *12*, 1411. [CrossRef]
11. Aziz, S.B.; Brza, M.A.; Hamsan, H.M.; Kadir, M.F.Z.; Abdulwahid, R.T. Electrochemical characteristics of solid-state double-layer capacitor constructed from proton conducting chitosan-based polymer blend electrolytes. *Polym. Bull.* **2020**. [CrossRef]
12. Aziz, S.B.; Hamsan, M.H.; Abdullah, R.M.; Abdulwahid, R.T.; Brza, M.A.; Marif, A.S.; Kadir, M.F.Z. Protonic EDLC cell based on chitosan (CS): Methylcellulose (MC) solid polymer blend electrolytes. *Ionics* **2020**, *26*, 1829–1840. [CrossRef]
13. Hwang, H.; Park, S.Y.; Kim, J.K.; Kim, Y.M.; Moon, H.C. Star-Shaped Block Copolymers: Effective Polymer Gelators of High-Performance Gel Electrolytes for Electrochemical Devices. *ACS Appl. Mater. Interfaces* **2019**, *11*, 4399–4407. [CrossRef]
14. Singh, P.; Bharati, D.C.; Gupta, P.N.; Saroj, A. Vibrational, thermal and ion transport properties of PVA-PVP-PEG-MeSO4Na based polymer blend electrolyte films. *J. Non-Cryst. Solids* **2018**, *494*, 21–30. [CrossRef]
15. Prajapati, G.; Roshan, R.; Gupta, P.N. Effect of plasticizer on ionic transport and dielectric properties of PVA–H3PO4 proton conducting polymeric electrolytes. *J. Phys. Chem. Solids* **2010**, *71*, 1717–1723. [CrossRef]
16. Misenan, M.S.M.; Isa, M.I.N.M.; Khiar, A.S.A. Electrical and structural studies of polymer electrolyte based on chitosan/methyl cellulose blend doped with BMIMTFSI. *Mater. Res. Express* **2018**, *5*, 055304. [CrossRef]
17. Radha, K.P.; Selvasekarapandian, S.; Karthikeyan, S.; Hema, M.; Sanjeeviraja, C. Synthesis and impedance analysis of proton-conducting polymer electrolyte PVA: NH4F. *Ionics* **2013**, *19*, 1437–1447. [CrossRef]
18. Aziz, S.B.; Hamsan, M.H.; Karim, W.O.; Kadir, M.; Brza, M.A.; Abdullah, O.G. High Proton Conducting Polymer Blend Electrolytes Based on Chitosan:Dextran with Constant Specific Capacitance and Energy Density. *Biomolecules* **2019**, *9*, 267. [CrossRef] [PubMed]
19. Lee, D.-Y.; Sohn, J.I.; Ahn, H.-J. High-surface-area tofu based activated porous carbon for electrical double-layer capacitors. *J. Ind. Eng. Chem.* **2017**, *52*, 121–127. [CrossRef]
20. Guo, J.; Jiang, J.; Yang, B. Low-voltage electric-double-layer MoS2 transistor gated via water solution. *Solid-State Electron.* **2018**, *150*, 8–15. [CrossRef]
21. Andres, B.; Dahlström, C.; Blomquist, N.; Norgren, M.; Olin, H. Cellulose binders for electric double-layer capacitor electrodes: The influence of cellulose quality on electrical properties. *Mater. Des.* **2018**, *141*, 342–349. [CrossRef]
22. Yang, I.; Kim, S.-G.; Kwon, S.H.; Lee, J.H.; Kim, M.-S.; Jung, J.C. Pore size-controlled carbon aerogels for EDLC electrodes in organic electrolytes. *Curr. Appl. Phys.* **2016**, *16*, 665–672. [CrossRef]
23. Tran, C.; Kalra, V. Fabrication of porous carbon nanofibers with adjustable pore sizes as electrodes for supercapacitors. *J. Power Sour.* **2013**, *235*, 289–296. [CrossRef]
24. Zhao, X.-Y. Key Laboratory of Coal Processing and Efficient Utilization (Ministry of Education), China University of Mining & Technology, Xuzhou 221116, Jiangsu, China Preparation and Characterization of Activated Carbons from Oxygen-rich Lignite for Electric Double-layer Capacitor. *Int. J. Electrochem. Sci.* **2018**, *13*, 2800–2816. [CrossRef]
25. Xu, M.; Li, D.; Yan, Y.; Guo, T.; Pang, H.; Xue, H. Porous high specific surface area-activated carbon with co-doping N, S and P for high-performance supercapacitors. *RSC Adv.* **2017**, *7*, 43780–43788. [CrossRef]
26. Nurhaziqah, A.M.S.; Afiqah, I.Q.; Aziz, M.F.H.A.; Aziz, N.A.N.; Hasiah, S. Optical, Structural and Electrical Studies of Biopolymer Electrolytes Based on Methylcellulose Doped with Ca (NO3)2. *IOP Conf. Series Mater. Sci. Eng.* **2018**, *440*, 012034. [CrossRef]
27. Aziz, S.B.; Hamsan, M.H.; Abdullah, R.M.; Kadir, M.F.Z. A Promising Polymer Blend Electrolytes Based on Chitosan: Methyl Cellulose for EDLC Application with High Specific Capacitance and Energy Density. *Molecules* **2019**, *24*, 2503. [CrossRef]
28. Aziz, S.B.; Abidin, Z.H.Z.; Kadir, M.F.Z. Innovative method to avoid the reduction of silver ions to silver nanoparticles ($Ag^+ \rightarrow Ag^o$) in silver ion conducting based polymer electrolytes. *Phys. Scr.* **2015**, *90*, 35808. [CrossRef]
29. He, Y.; Zhu, B.; Inoue, Y. Hydrogen bonds in polymer blends. *Prog. Polym. Sci.* **2004**, *29*, 1021–1051. [CrossRef]

30. Lu, G.; Kong, L.; Sheng, B.; Wang, X.; Gong, Y.; Zhang, X. Degradation of covalently cross-linked carboxymethyl chitosan and its potential application for peripheral nerve regeneration. *Eur. Polym. J.* **2007**, *43*, 3807–3818. [CrossRef]
31. Aziz, S.B.; Marif, R.B.; Brza, M.; Hassan, A.N.; Ahmad, H.A.; Faidhalla, Y.A.; Kadir, M. Structural, thermal, morphological, and optical properties of PEO filled with biosynthesized Ag nanoparticles: New insights to band gap study. *Results Phys.* **2019**, *13*, 102220. [CrossRef]
32. Wen, S.; Richardson, T.; Ghantous, D.; Striebel, K.; Ross, P.; Cairns, E.J. FTIR characterization of PEO + LiN (CF3SO2)2 electrolytes. *J. Electroanal. Chem.* **1996**, *408*, 113–118. [CrossRef]
33. Aziz, S.B.; Abidin, Z.H.Z. Electrical Conduction Mechanism in Solid Polymer Electrolytes: New Concepts to Arrhenius Equation. *J. Soft Matter* **2013**, 1–8. [CrossRef]
34. Sahli, N.B.; Ali, A.M.M. Effect of lithium triflate salt concentration in methyl cellulose-based solid polymer electrolytes. In Proceedings of the 2012 IEEE Colloquium on Humanities, Science and Engineering (CHUSER), Kota Kinabalu, Malaysia, 3–4 December 2012; pp. 739–742. [CrossRef]
35. Kamarudin, K.H.; Isa, M.I.N. Structural and DC Ionic conductivity studies of carboxy methylcellulose doped with ammonium nitrate as solid polymer electrolytes. *Int. J. Phys. Sci.* **2013**, *8*, 1581–1587. [CrossRef]
36. Samsudin, A.S.; Kuan, E.C.H.; Isa, M.I.N.M. Investigation of the Potential of Proton-Conducting Biopolymer Electrolytes Based Methyl Cellulose-Glycolic Acid. *Int. J. Polym. Anal. Charact.* **2011**, *16*, 477–485. [CrossRef]
37. Aziz, S.B.; Karim, W.O.; Qadir, K.; Zafar, Q. Proton Ion Conducting Solid Polymer Electrolytes Based on Chitosan Incorporated with Various Amounts of Barium Titanate (BaTiO3). *Int. J. Electrochem. Sci.* **2018**, *13*, 6112–6125. [CrossRef]
38. Pradhan, D.K.; Choudhary, R.N.; Samantaray, B.K.; Karan, N.K.; Katiyar, R.S. Effect of Plasticizer on Structural and Electrical Properties of Polymer Nanocompsoite Electrolytes. *Int. J. Electrochem. Sci.* **2007**, *2*, 861–871.
39. Mohapatra, S.R.; Thakur, A.K.; Choudhary, R. Effect of nanoscopic confinement on improvement in ion conduction and stability properties of an intercalated polymer nanocomposite electrolyte for energy storage applications. *J. Power Sour.* **2009**, *191*, 601–613. [CrossRef]
40. Aziz, S.B.; Abdullah, R.M.; Kadir, M.; Ahmed, H.M. Non suitability of silver ion conducting polymer electrolytes based on chitosan mediated by barium titanate (BaTiO3) for electrochemical device applications. *Electrochim. Acta* **2019**, *296*, 494–507. [CrossRef]
41. Aziz, S.B.; Woo, T.J.; Kadir, M.F.; Ahmed, H.M.; Ahmed, H.M. A conceptual review on polymer electrolytes and ion transport models. *J. Sci. Adv. Mater. Devices* **2018**, *3*, 1–17. [CrossRef]
42. Aziz, S.B.; Hamsan, M.; Brza, M.; Kadir, M.; Abdulwahid, R.T.; Ghareeb, H.O.; Woo, H. Fabrication of energy storage EDLC device based on CS: PEO polymer blend electrolytes with high Li+ ion transference number. *Results Phys.* **2019**, *15*, 102584. [CrossRef]
43. Sekhon, S.S.; Agnihotry, S.A. *Solid State Ionics: Science and Technology*; Chowdari, B.V.R., Lal, K., Agnihotry, S.A., Khare, N., Sekhon, S.S., Srivastava, P.C., Chandra, S., Eds.; World Scientific: Singapore, 1998; p. 527.
44. Kufian, M.Z.; Aziz, M.F.; Shukur, M.; Rahim, A.; Ariffin, N.; Shuhaimi, N.; Majid, S.; Yahya, R.; Arof, A.K. PMMA–LiBOB gel electrolyte for application in lithium ion batteries. *Solid State Ionics* **2012**, *208*, 36–42. [CrossRef]
45. Ramlli, M.A.; Isa, M.I.N.M. Structural and Ionic Transport Properties of Protonic Conducting Solid Biopolymer Electrolytes Based on Carboxymethyl Cellulose Doped with Ammonium Fluoride. *J. Phys. Chem. B* **2016**, *120*, 11567–11573. [CrossRef] [PubMed]
46. Diederichsen, K.M.; McShane, E.J.; McCloskey, B.D. Promising Routes to a High Li+ Transference Number Electrolyte for Lithium Ion Batteries. *ACS Energy Lett.* **2017**, *2*, 2563–2575. [CrossRef]
47. Pratap, R.; Singh, B.; Chandra, S. Polymeric rechargeable solid-state proton battery. *J. Power Sour.* **2006**, *161*, 702–706. [CrossRef]
48. Sampathkumar, L.; Selvin, P.C.; Selvasekarapandian, S.; Perumal, P.; Chitra, R.; Muthukrishnan, M. Synthesis and characterization of biopolymer electrolyte based on tamarind seed polysaccharide, lithium perchlorate and ethylene carbonate for electrochemical applications. *Ionics* **2019**, *25*, 1067–1082. [CrossRef]
49. Aziz, S.B.; Hamsan, M.H.; Kadir, M.F.Z.; Karim, W.O.; Abdullah, R.M. Development of Polymer Blend Electrolyte Membranes Based on Chitosan: Dextran with High Ion Transport Properties for EDLC Application. *Int. J. Mol. Sci.* **2019**, *20*, 3369. [CrossRef]

50. Shuhaimi, N.E.A.; Teo, L.P.; Woo, H.J.; Majid, S.R.; Arof, A.K. Electrical double-layer capacitors with plasticized polymer electrolyte based on methyl cellulose. *Polym. Bull.* **2012**, *69*, 807–826. [CrossRef]
51. Virya, A.; Lian, K. Lithium polyacrylate-polyacrylamide blend as polymer electrolytes for solid-state electrochemical capacitors. *Electrochem. Commun.* **2018**, *97*, 77–81. [CrossRef]
52. Kadir, M.F.Z.; Arof, A.K. Application of PVA–chitosan blend polymer electrolyte membrane in electrical double layer capacitor. *Mater. Res. Innov.* **2011**, *15*, s217–s220. [CrossRef]
53. Lim, C.-S.; Teoh, K.H.; Liew, C.W.; Ramesh, S. Electric double layer capacitor based on activated carbon electrode and biodegradable composite polymer electrolyte. *Ionics* **2013**, *20*, 251–258. [CrossRef]
54. Liew, C.W.; Ramesh, S.; Arof, A.K. Enhanced capacitance of EDLCs (electrical double layer capacitors) based on ionic liquid-added polymer electrolytes. *Energy* **2016**, *109*, 546–556. [CrossRef]
55. Aziz, S.B.; Abdulwahid, R.T.; Hamsan, M.H.; Brza, M.A.; Abdullah, R.M.; Kadir, M.; Muzakir, S.K. Structural, Impedance, and EDLC Characteristics of Proton Conducting Chitosan-Based Polymer Blend Electrolytes with High Electrochemical Stability. *Molecules* **2019**, *24*, 3508. [CrossRef] [PubMed]
56. Aziz, S.B.; Brza, M.; Hamsan, M.; Kadir, M.; Muzakir, S.; Abdulwahid, R.T. Effect of ohmic-drop on electrochemical performance of EDLC fabricated from PVA: Dextran: NH4I based polymer blend electrolytes. *J. Mater. Res. Technol.* **2020**, *9*, 3734–3745. [CrossRef]
57. Selvakumar, M.; Bhat, D.K. Polyvinyl alcohol–polystyrene sulphonic acid blend electrolyte for supercapacitor application. *Phys. B Condens. Matter* **2009**, *404*, 1143–1147. [CrossRef]
58. Shukur, M.; Ithnin, R.; Kadir, M. Protonic Transport Analysis of Starch-Chitosan Blend Based Electrolytes and Application in Electrochemical Device. *Mol. Cryst. Liq. Cryst.* **2014**, *603*, 52–65. [CrossRef]
59. Arof, A.K.; Kufian, M.Z.; Shukur, M.; Aziz, M.F.; Abdelrahman, A.; Majid, S. Electrical double layer capacitor using poly (methyl methacrylate)–C4BO8Li gel polymer electrolyte and carbonaceous material from shells of mata kucing (Dimocarpus longan) fruit. *Electrochim. Acta* **2012**, *74*, 39–45. [CrossRef]
60. Yang, H.; Kannappan, S.; Pandian, A.S.; Jang, J.-H.; Lee, Y.S.; Lu, W. Graphene supercapacitor with both high power and energy density. *Nanotechnology* **2017**, *28*, 445401. [CrossRef]
61. Kadir, M.; Hamsan, M.H. Green electrolytes based on dextran-chitosan blend and the effect of NH4SCN as proton provider on the electrical response studies. *Ionics* **2017**, *24*, 2379–2398. [CrossRef]
62. Hamsan, M.H.; Shukur, M.; Kadir, M. The effect of NH4NO3 towards the conductivity enhancement and electrical behavior in methyl cellulose-starch blend based ionic conductors. *Ionics* **2016**, *23*, 1137–1154. [CrossRef]
63. Pesko, D.M.; Jung, Y.; Hasan, A.L.; Webb, M.A.; Coates, G.W.; Miller, T.F.; Balsara, N.P. Effect of monomer structure on ionic conductivity in a systematic set of polyester electrolytes. *Solid State Ionics* **2016**, *289*, 118–124. [CrossRef]
64. Hamsan, M.H.; Shukur, M.; Kadir, M. NH4NO3 as charge carrier contributor in glycerolized potato starch-methyl cellulose blend-based polymer electrolyte and the application in electrochemical double-layer capacitor. *Ionics* **2017**, *23*, 3429–3453. [CrossRef]
65. Liew, C.W.; Ramesh, S. Electrical, structural, thermal, and electrochemical properties of corn starch-based biopolymer electrolytes. *Carbohydr. Polym.* **2015**, *124*, 222–228. [CrossRef]
66. Asnawi, A.S.F.M.; Aziz, S.B.; Nofal, M.M.; Yusof, Y.M.; Brevik, I.; Hamsan, M.H.; Brza, M.A.; Abdilwahid, R.; Kadir, M. Metal Complex as a Novel Approach to Enhance the Amorphous Phase and Improve the EDLC Performance of Plasticized Proton Conducting Chitosan-Based Polymer Electrolyte. *Membranes* **2020**, *10*, 132. [CrossRef] [PubMed]
67. Asnawi, A.S.F.M.; Aziz, S.B.; Nofal, M.M.; Hamsan, M.H.; Brza, M.; Yusof, Y.M.; Abdilwahid, R.T.; Muzakir, S.K.; Kadir, M.F.Z. Glycerolized Li^+ Ion Conducting Chitosan-Based Polymer Electrolyte for Energy Storage EDLC Device Applications with Relatively High Energy Density. *Polymers* **2020**, *12*, 1433. [CrossRef] [PubMed]
68. Hamsan, M.; Aziz, S.B.; Azha, M.; Azli, A.; Shukur, M.; Yusof, Y.; Muzakir, S.; Manan, N.S.; Kadir, M. Solid-state double layer capacitors and protonic cell fabricated with dextran from Leuconostoc mesenteroides based green polymer electrolyte. *Mater. Chem. Phys.* **2020**, *241*, 122290. [CrossRef]

69. Aziz, S.B.; Brza, M.; Mishra, K.; Hamsan, M.; Karim, W.O.; Abdullah, R.M.; Kadir, M.; Abdulwahid, R.T. Fabrication of high-performance energy storage EDLC device from proton conducting methylcellulose: Dextran polymer blend electrolytes. *J. Mater. Res. Technol.* **2020**, *9*, 1137–1150. [CrossRef]
70. Aziz, S.B.; Hamsan, M.H.; Karim, W.O.; Marif, A.S.; Abdulwahid, R.T.; Kadir, M.F.Z.; Brza, M.A. Study of impedance and solid-state double-layer capacitor behavior of proton (H+)-conducting polymer blend electrolyte-based CS:PS polymers. *Ionics* **2020**, 1–15. [CrossRef]

Article

A Deep Insight into Different Acidic Additives as Doping Agents for Enhancing Proton Conductivity on Polybenzimidazole Membranes

Jorge Escorihuela [1,*], Abel García-Bernabé [2] and Vicente Compañ [2,*]

[1] Departamento de Química Orgánica, Facultad de Farmacia, Universitat de València, Av. Vicent Andrés Estellés s/n, 46100 Burjassot, Valencia, Spain

[2] Departamento de Termodinámica Aplicada, Escuela Técnica Superior de Ingeniería Industrial, Universitat Politècnica de València, Camino de Vera s/n, 46022 Valencia, Spain; agarciab@ter.upv.es

* Correspondence: jorge.escorihuela@uv.es (J.E.); vicommo@ter.upv.es (V.C.); Tel.: +34-96-387-9328 (V.C.)

Received: 27 May 2020; Accepted: 15 June 2020; Published: 18 June 2020 

Abstract: The use of phosphoric acid doped polybenzimidazole (PBI) membranes for fuel cell applications has been extensively studied in the past decades. In this article, we present a systematic study of the physicochemical properties and proton conductivity of PBI membranes doped with the commonly used phosphoric acid at different concentrations (0.1, 1, and 14 M), and with other alternative acids such as phytic acid (0.075 M) and phosphotungstic acid (HPW, 0.1 M). The use of these three acids was reflected in the formation of channels in the polymeric network as observed by cross-section SEM images. The acid doping enhanced proton conductivity of PBI membranes and, after doping, these conducting materials maintained their mechanical properties and thermal stability for their application as proton exchange membrane fuel cells, capable of operating at intermediate or high temperatures. Under doping with similar acidic concentrations, membranes with phytic acid displayed a superior conducting behavior when compared to doping with phosphoric acid or phosphotungstic acid.

Keywords: fuel cells; proton conductivity; electrochemical impedance spectroscopy; polymer; polybenzimidazole; proton exchange membrane; phosphoric acid; phytic acid; phosphotungstic acid

---

1. Introduction

Carbon dioxide concentration in the atmosphere has reached worrying values above 400 ppm in the last months (Figure 1) [1]. These alarming values have focused researchers' objectives towards the search for novel and sustainable energy systems to fulfil the world demand [2]. Fuel cells are electrochemical devices that convert chemical energy into electrical energy in an efficient way [3]. Among the different types of fuel cell, proton exchange membrane fuel cells (PEMFCs), which use an ion exchange polymer film as the electrolyte, have received special attention due to their low operating temperature and quick start-up [4,5]. The main component of a PEMFC is the proton exchange membrane (PEM), which is responsible for proton transport between the anode and the cathode [6]. In this regard, the synthesis of novel polymer electrolyte materials with high chemical, thermal and mechanical stability combined with elevated conductivity, has absorbed the market over the past few decades [7–9]. Among the wide array of polymers used for this purpose, Nafion is the most commonly used PEM material and possesses high proton conductivity as well as excellent chemical stability at low temperatures [10]; however, its conductivity drops at temperatures above 90 °C, hampering its use as a high temperature PEMFC [11]. The advantages of working at temperatures above 120 °C include a remarkable reduction in CO poisoning of the catalyst [12], improvement of diffusion rates and redox reactions [13], and simplification in water and heat management [14]. In the quest for novel polymeric electrolytes for energy applications operating at high temperatures [15], several polymeric families have been developed in order to replace the widely used perfluorinated polymers [16–18]. Among these novel polymeric materials, polybenzimidazole (PBI) derivatives have emerged as potential candidates due to their higher thermal and mechanical stability [19].

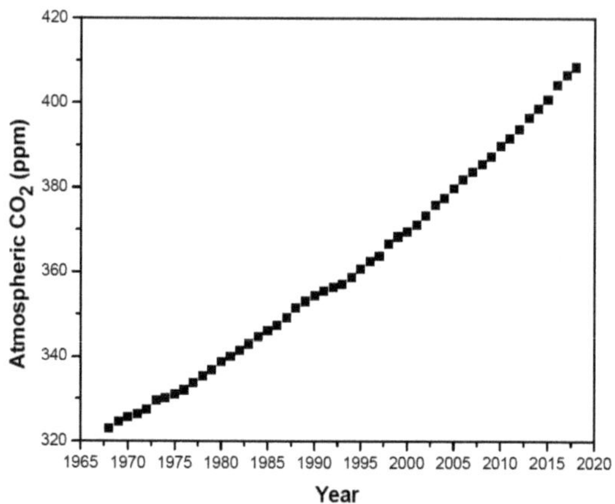

Figure 1. Atmospheric CO_2 concentration (ppm) in the past 25 years.

PBI (poly[2,2′-(m-phenylen)-5,5′-bisbenzimidazole]) is a heterocyclic, thermoplastic, basic, and hydrophilic synthetic polymer with a very high glass transition temperature (425–436 °C) [20]. Since its development in 1961 by H. Vogel and C.S. Marvel [21], PBI has been used by NASA in the Apollo missions as part of the astronauts' clothing, but it was not until 1995 that it was used in fuel cells by Wainright et al. [22]. Despite possessing high and thermal chemical stability, combined with good mechanical stability, the proton conductivity of PBI membranes is relatively low without chemical modification of the PBI structure or the introduction of fillers into the PBI matrix. In this regard, the most widely used strategy to increase proton conductivity in PBI-based membranes is acid doping. In particular, the use of phosphoric acid (PA) has been established as a standard approach [23–27], despite drawbacks such as acid leaching and membrane degradation [28,29]. Throughout the past decades, a variety of alternatives to PA have been used in an attempt to improve the physicochemical properties and performance of PBI membranes, such as the incorporation of inorganic fillers [30,31], metallic salts [32,33], carbon-based materials [34,35], zeolitic imidazolate frameworks (ZIFs) [36–38], and ionic liquids (ILs), among others [39–42].

As mentioned above, phosphoric acid (PA) is by far the most extensively used approach to enhancing proton conductivity of PBI-based membranes. After acid doping, conductivity can be enhanced by several orders of magnitude, reaching values in the range of 0.1–0.2 $S{\cdot}cm^{-1}$, which are comparable to Nafion membranes operating at 80 °C under high hydration conditions. However, acid leaching studies and stability tests after a few operating cycles are generally omitted in these studies. Among other doping acids used to increase proton conductivity in polymer electrolyte membranes, phytic acid (myoinositol hexakisphosphate) and phosphotungstic acid (HPW) are among the most prominent and have emerged as more sustainable alternatives to replace phosphoric acid (Figure 2). Phytic acid is a phosphorus-containing organic acid that is present in plants, especially in seeds and fiber. This acid has been used as a doping agent for polymer electrolyte membranes based on Nafion, yielding excellent proton conductivities [43,44]. On the other hand, HPW, an heteropoly acid with molecular formula $H_3P_4W_{12}O_{40}$, has also been efficiently applied as a proton carrier in proton exchange membrane fuel cells [45–48].

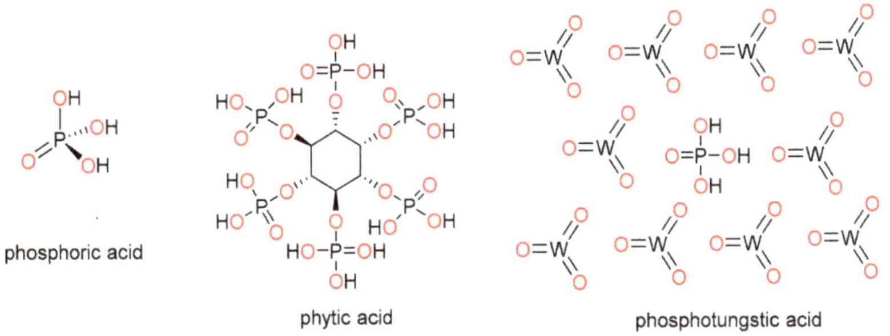

phosphoric acid

phytic acid

phosphotungstic acid

Figure 2. Chemical structures of phosphoric acid (PA), phytic acid, and phosphotungstic acid (HPW).

Continuing our ongoing work towards developing novel proton exchange membranes based on polybenzimidazole for high temperature PEMFCs, we present a systematic study of physicochemical properties and proton conductivity of PBI membranes doped with phosphoric acid at different concentrations (0.1, 1, and 14 M) and other alternative acids, such as phytic acid (0.075 M) and HPW (0.1 M). The use of these three acids was reflected in the formation of channels in the polymeric network as observed by cross-section SEM images. These membranes exhibited improved proton conductivity compared to undoped PBI membranes. We also studied other properties of the PBI-doped membranes, including their phosphoric acid uptake, thermal stability, and mechanical strength. Finally, we also calculated the proton diffusion coefficient (*D*) using the Nernst–Einstein equation.

2. Materials and Methods

2.1. Materials

PBI (purity >99.95%, molecular formula: $(C_{20}H_{12}N_4)_n$, MW 51,000,) was purchased from Danish Power Systems. Lithium chloride (LiCl), *N,N*-dimethylacetamide (DMAc) 99.8%, phosphoric acid (extra pure, 85% solution in water), phytic acid solution (50% (*w/w*) in H_2O), sodium hydrogen phosphate (Na_2HPO_4), and sodium tungstate dihydrate ($Na_2WO_4{\cdot}2H_2O$) were purchased from Sigma-Aldrich (Aldrich, Madrid, Spain). The phosphotungstic acid solution was prepared using $Na_2WO_4{\cdot}2H_2O$ and Na_2HPO_4 as precursors, by mixing 1.10 g of Na_2HPO_4 and 5.95 g of $Na_2WO_4{\cdot}2H_2O$ in 30 mL of deionized water at 50 °C.

2.2. Membrane Preparation.

Initially, a DMAc solution containing 0.1 wt.% of LiCl (used as a stabilizer) was prepared by dissolving 100 mg of LiCl in 100 mL of DMAc under vigorous stirring (1 h at room temperature) to give a homogeneous solution. Next, a 10 wt.% PBI solution was prepared by dissolving PBI powder (10 g) in the DMAc solution (90 g). The mixture was heated under reflux at 120 °C for 6 h. The final prepared solution had a viscosity of 0.5 Pa·s at 25 °C. Then, PBI membranes were prepared via the solution cast method. To this end, the PBI solution was cast onto a clean glass slide and dried under vacuum at 80 °C for 16 h and finally at 140 °C for 10 h. Membranes were washed with distilled water at 80 °C in order to remove residual solvent (DMAc) and LiCl. Traces of the solvent were finally removed by drying at 160 °C for 16 h. The membrane thicknesses prior to acid doping varied between 100 and 120 μm. Finally, membranes were doped with the corresponding acid by immersion in the acidic solution for 48 h at room temperature.

2.3. Membrane Characterization

Scanning electron microscopy (SEM) images were obtained using a field emission scanning electron microscope (FE-SEM) model Ultra 55 (Zeiss) operating at 5 kV with energy-dispersive X-ray (EDX) spectroscopy. Attenuated total reflection Fourier transform infrared (ATR-FTIR) spectra of the membranes were obtained using a Jasco FTIR spectrometer FT/IR-6200 Series (Jasco) with a 4 cm^{-1} resolution between 4000 and 600 cm^{-1}. Thermogravimetric analysis (TGA) was performed on a TGA Q50 thermogravimetric analyzer TGA Q50 (Waters) under nitrogen atmosphere (60 mL·min^{-1}) from 30 to 800 °C using a heating rate of 10 °C·min^{-1}. The acid uptake (AU) of the membrane was calculated by the following equation: AU (%) = $[(W_{wet} - W_{dry})/W_{dry}] \times 100$; where W_{wet} and W_{dry} refer to the membrane's weight after its immersion in the acid solution for at least 48 h at room temperature and the membrane's weight after drying at 110 °C for at least 24 h, respectively. The thickness uptake (TU) of the membrane was calculated by the following equation: TU (%) = $[(T_{wet} - T_{dry})/W_{dry}] \times 100$; where T_{wet} and T_{dry} refer to the membrane's thickness weight after drying at 110 °C for 24 h, respectively. The tensile properties of the membranes were calculated from stress–strain curves obtained using a precision universal/tensile tester (Shimadzu AGS-X) at a crosshead rate of 10 mm·min^{-1} at room temperature. To this end, membranes (five samples of each membrane) with a thickness around 100 μm were cut into strips of 30 mm × 6 mm and tested. Proton conductivity measurements (in the transversal direction) were performed using a broadband dielectric spectrometer (Novocontrol Technologies, Montabaur, Germany) integrated with an SR 830 lock-in amplifier with an alpha dielectric interface from 20 to 200 °C by electrochemical impedance spectroscopy (EIS) in the frequency interval of 0.1 Hz to 10 MHz, applying a 0.1 V signal amplitude. Initially, the temperature was gradually raised from 20 to 120 °C in steps of 20 °C and the dielectric spectra were collected at each step. During the measurements, the temperature was isothermally controlled using a nitrogen jet (Quatro from Novocontrol, Montabaur, Germany) with a temperature error of 0.1 °C during every single sweep in frequency.

3. Results and Discussion

3.1. Membrane Preparation and Characterization

PBI membranes were prepared by the casting method (Figure 3). For this purpose, a 10 wt.% PBI solution was prepared using a DMAc solution containing 0.1 wt.% of LiCl (used as a stabilizer). Shortly, PBI powder was completely dissolved in the DMAc solution, and then the homogeneous solution was cast onto a clean glass plate and heated in a vacuum oven at 80 °C for 16 h to completely remove the DMAc, and finally heated at 140 °C for 10 h. Using this methodology, transparent membranes with a uniform thickness of around 100 μm were obtained.

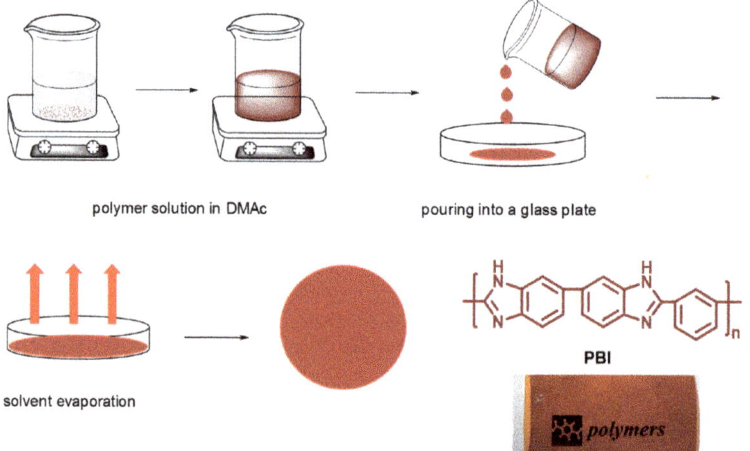

Figure 3. Schematic representation of polybenzimidazole (PBI) membrane preparation by the casting method.

Next, PBI membranes were doped with phosphoric acid (0.1, 1, and 14 M), phytic acid (0.075 M), and HPW (0.1 M). Specifically, the acid-doped membranes were obtained by soaking the membranes in the corresponding acid solution at room temperature for 2 days to ensure complete saturation in the membrane. After this time, the membranes were removed, wiped down, and dried in a vacuum oven at 120 °C for 48 h to obtain a constant weight. The acid doping was confirmed by FTIR by the presence of a broad band centered at 1000 cm^{-1} corresponding to phosphonate groups (see Figure S1). FTIR spectra of membranes doped with phosphoric and phytic acid, displayed this intense characteristic band; however, it was very low for the phosphotungstic acid membrane, indicating a low degree of doping of this heteropoly acid in the PBI–HPW membrane.

Acid uptake and swelling are critical parameters to be considered when studying polymeric membranes for PEM fuel cell applications [49]. In this regard, the performance of a membrane is generally evaluated according to its proton conductivity, which is strongly dependent on its water or acid content. In this regard, high proton conductivity is associated with high levels of acid uptake; at the same time, it is also a sign of low-dimensional stability, as acid modifies the polymer microstructure and mechanical properties. Table 1 shows the acid uptake (AU), swelling, and thickness uptake (TU) of the PBI membranes measured at room temperature after immersion of 48 h in deionized water (DIW), phosphoric acid at different concentrations (0.1, 1, and 14 M), and phytic acid and HPW, at 0.075 and 0.1 M, respectively. The acid doping level was calculated by measuring the weight changes between dry doped membranes (W_{acid}) and dry undoped membranes (W_{dry}), where M_{acid} and M_{PBI} represent the molecular weight of acid (97.99 for PA, 660.04 for phytic acid, and 2880.05 for HPW) and the repeating unit of PBI (308.34 for $C_{20}H_{12}N_4$), respectively.

$$\text{acid doping level} = \frac{(W_{acid} - W_{dry})/M_{acid}}{W_{dry}/M_{PBI}} \tag{1}$$

As mentioned above, the acid doping level is a fundamental parameter for proton transport and expresses the number of acid molecules per monomer unit in the polymeric network. Accordingly, the higher the acid doping level of the membrane, the higher its proton conductivity. As shown in Table 1, the acid doping level was dependent on acid concentration and, as expected, higher acid doping levels were obtained for the membrane doped using the highest concentration (PBI–PA 14 M). Accordingly, higher swelling and TU were also observed for higher acid concentrations. In contrast, low acid doping levels were obtained when using phytic acid and HPW as alternative acids to conventional phosphoric acid doping, as also observed in previous studies by Kawakami and coworkers [44]. In this regard, to have a fair comparison, acid doping levels need to be compared with PBI–PA 0.1 M, as phytic acid and HPW were used at concentrations of 0.075 and 0.1 M, respectively. Due to the low solubility of phytic acid in water, the concentration of phytic acid corresponds to the commercially available phytic acid solution (50% (*w/w*) in H_2O).

Table 1. Acid uptake (AU), swelling, and thickness uptake (TU) of the PBI membranes doped at room temperature under different conditions. DIW, deionized water.

Membrane	AU (%)	Swelling (%)	TU (%)	Acid Doping Level (%)
PBI–DIW	4 ± 1	5 ± 1	4 ± 1	—
PBI–PA 0.1 M	19 ± 1	9 ± 1	8 ± 1	0.60
PBI–PA 1 M	63 ± 3	17 ± 1	22 ± 2	1.98
PBI–PA 14 M	278 ± 5	63 ± 4	80 ± 3	8.74
PBI–phytic acid	21 ± 1	12 ± 1	10 ± 1	0.10
PBI–HPW	25 ± 1	15 ± 1	9 ± 1	0.03

The internal microscopic morphologies of membranes were studied by SEM images. The SEM images of cryo-fractured cross-sections of the different PBI membranes are shown in Figure 4. The cross-section morphology of the undoped PBI membrane was dense and free of holes. However, after the addition of PA, the morphology of all membranes showed the formation of channels due to the presence of the acidic filler, reflected in the appearance of holes in the cross-section SEM images. After PA doping, the morphology of all membranes showed the formation of channels in the polymer network due to the presence of PA molecules, as observed in similar systems [50]. The doping with phytic acid and HPW also showed the presence of microstructures which might be involved in the conduction process. It could also be observed that the rough cross-section microscopic morphologies of PA-doped membranes resulted in a rough fracture cross-section attributed to the plasticizing effect of phosphoric acid.

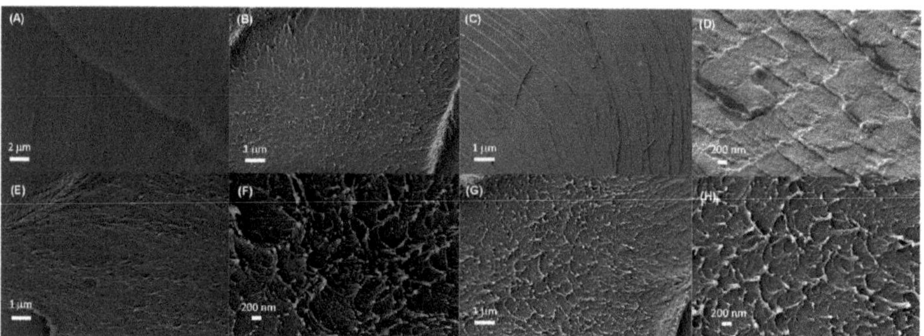

Figure 4. SEM images of cross-sections of PBI membranes after immersion in DIW (**A**,**B**), phosphoric acid 0.1 M (**C**,**D**), phytic acid 0.075 M (**E**,**F**), and HPW 0.1 M (**G**,**H**).

One of the major properties to be considered in the development of proton exchange membrane for fuel cell applications is their thermal stability [51]. The thermal behavior of the membranes was studied via thermal analysis under N_2 atmosphere with a 10 °C·min^{-1} heating rate. As shown in Figure 5, the thermal stability of PBI membranes was investigated by thermogravimetric analysis (TGA). For the PA-doped membranes, all the curves displayed a similar trend. The first degradation step was observed at 160 °C, which is attributed to the formation of pyrophosphoric acid through a condensation reaction of phosphoric acid. A second step was observed at about 600 °C and was associated to the degradation of the PBI main chain and the continuous dehydration of the pyrophosphoric acid to polyphosphoric acid. The results are similar to previously reported PBI membranes [46]. The weight loss curves for the PBI–phytic acid and PBI–HPW membranes presented a similar degradation trend. Furthermore, all the membranes showed similar $T_{d5\%}$ to the pristine PBI membrane. From the thermogravimetric analysis it can be concluded that the reported membranes possess an adequate thermal stability for their application as proton exchange membrane fuel cells for intermediate or high temperatures.

The study of mechanical properties of polymer electrolyte membranes is of the utmost importance for future application as PEM fuel cells [52]. In this regard, PEMs possessing excellent mechanical properties are highly demanded; however, high values of PA doping generally produce a decrease in the mechanical strength of PBI membranes, as PA molecules reduce the interaction between polymeric chains. The tensile properties of the acid-doped membranes were determined from stress–strain curves obtained with a universal testing machine at a crosshead rate of 10 mm·min^{-1} at room temperature. For that, rectangular samples (30 mm × 6 mm) with a thickness of 150 µm (five samples of each type of membrane) were tested and the average results are given in Table 2 with the corresponding standard deviation. For a better comparison, the Young's modulus, tensile stress, and elongation at break values of the pure PBI dry membrane are also included. The undoped pristine PBI membrane had a Young's modulus of 2.52 GPa, a tensile strength of 174 MPa, and an elongation at break of 18%. As observed, the immersion in water decreased both Young's modulus and tensile stress. Meanwhile, the value of elongation at break increased. Phosphoric acid doping produced a similar effect, being more marked as the acid concentration increased, due to the plasticizing effect of phosphoric acid. The mechanical properties of the phosphoric acid-doped PBI membranes can be improved by lowering the acid doping level; however, the proton conductivity is dramatically reduced. This significant reduction of the PA-doped PBI membrane strength has been reported by other researchers in the literature [53,54]. PBI membranes doped with low concentrations of other acidic fillers, such as phytic acid and HPW, also displayed high values of elongation at break, in the range of 119–125%. The values are comparable with those of other reported PBI membranes [55–58]. It should be noted that all the composite membranes exhibited a tensile strength of above 2.0 MPa, which was strong enough for the fabrication of membrane electrode assemblies (MEAs) that could be evaluated in fuel cell performance tests [59,60].

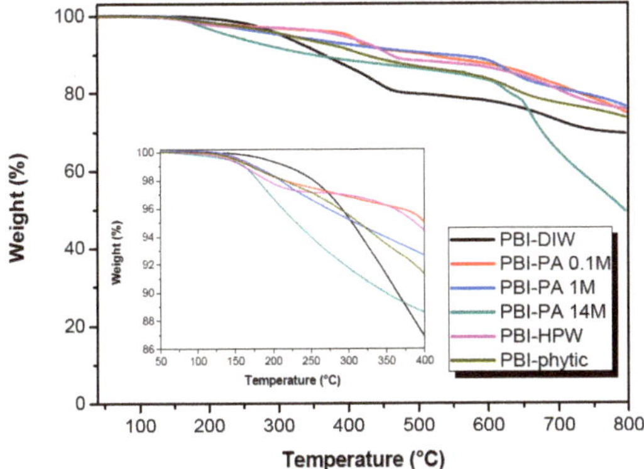

Figure 5. Thermogravimetric analysis of PBI membranes doped with different acids and concentrations. Inset: zoom at the 50–400 °C region.

Table 2. Mechanical properties of the PBI membranes doped under different conditions.

Membrane	Young's Modulus (GPa)	Tensile Stress (MPa)	Elongation at Break (%)
PBI–dry	2.52 ± 0.17	174 ± 4	18 ± 2
PBI–DIW	2.13 ± 0.21	98 ± 6	66 ± 3
PBI–PA 0.1 M	1.58 ± 0.19	92 ± 5	89 ± 5
PBI–PA 1 M	1.22 ± 0.12	75 ± 4	108 ± 7
PBI–PA 14 M	0.11 ± 0.02	19 ± 2	189 ± 9
PBI–phytic acid	1.11 ± 0.15	71 ± 2	119 ± 5
PBI–HPW	1.67 ± 0.13	90 ± 4	125 ± 6

3.2. Proton Conductivity

The analysis of proton conductivity is of the utmost importance for evaluation of a membrane to be considered as a membrane electrode assembly (MEA) in the manufacturing of proton exchange membrane fuel cells (PEMFCs). Among the different techniques generally used to measure proton conductivity in membranes, electrochemical impedance spectroscopy (EIS) has emerged as a powerful electrochemical technique to measure the proton conductivity of PEMFCs [61–63]. The through-plane conductivity of the PBI-doped membranes at different temperatures (from 20 to 200 °C) was determined by EIS using a blocking electrode configuration.

Generally, the proton conductivity in acid-doped membranes based on PBI are generally reported on a unique measurement at different temperatures. However, proton conductivity is not constant and generally drops after a few operating cycles. In order to evaluate the stability of proton conductivity in the acid-doped membranes, proton conductivity of the membranes was evaluated in four consecutive cycles. To this end, in the first ramp, measurements were performed in steps of 20 °C from 20 to 200 °C. Then, the sample was cooled down and a second ramp of measurements was applied in the temperature interval from 200 to 20 °C. This procedure was systematically repeated in the third (from

20 to 200 °C) and fourth ramp (200 to 20 °C) of measurements (see Figure S2). The dc-conductivity was obtained from the Bode diagram, which shows the real part of the conductivity (σ) as a function of the frequency. Ideally, the proton conductivity of the membrane can be obtained from the value of the conductivity in the region of high frequencies where the conductivity reaches a plateau [64]. Accordingly, the proton conductivity was obtained from the Bode plots in the different measurement cycles (Figure 6).

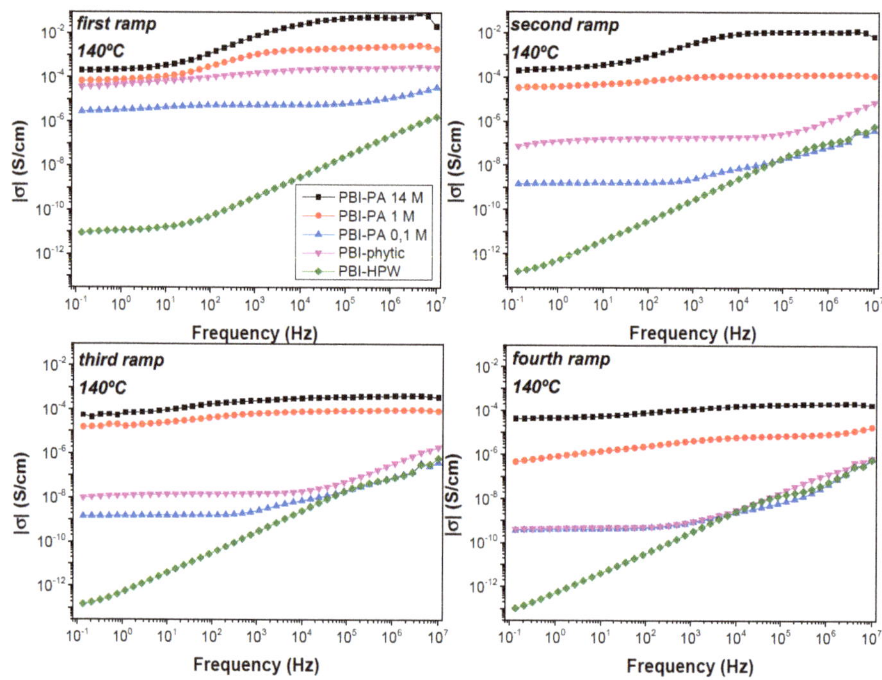

Figure 6. Bode diagrams at 140 °C for the acid-doped membranes: PBI–PA 0.1 M (▲), PBI–PA 1 M (•), PBI–PA 14 M (■),PBI–phytic acid (▼), and PBI–HPW (♦) measured in different cycles.

As shown in Figure 6, the values obtained for the dc-conductivity (σ) varied depending on the measurement cycle. Consequently, these results show that the conductivity of the membranes does not remain constant with temperature and time, which can be associated to the varying density of charge carrier into the membranes. In this regard, conductivity decreased along the different measurement cycles, changing about two orders of magnitude from the first to the fourth ramp in the case of PA-doped membranes PBI–PA 14 M and PBI–PA 1 M; around three orders of magnitude for the sample PBI–PA 0.1 M, and four orders of magnitude for the PBI–phytic acid membrane. This effect is attributed to the rapid loss of the free phosphoric acid molecules from the doped membranes. The leaching problems of acids from PBI membranes were evaluated in a long-term conductivity study of the membranes at 25 °C showing that a significant decrease in conductivity was observed after 24 h (Figure S3). This drawback was also observed in a long-term conductivity study at 10 °C (Figure 7), which is a usual operating temperature for high temperature (HT)-PEMFC membranes. As observed, and although PBI–PA 14 M membrane has a high proton conductivity, its value decreased around two orders of magnitude after 24 h, indicating an important acid leaching. Interestingly, PBI–phytic acid membrane has a lower leaching drawback and its conductivity at 160 °C was similar to that of PBI–PA 1 M membrane, contrary to what was observed at 25 °C, where PBI–PA 1 M membrane had a superior conductivity value. This result shows that phytic acid can be a promising candidate to be used at high temperatures. However, the use of other fillers such as ionic liquids and metal organic frameworks, among others, can help to more efficiently retain the acidic additive.

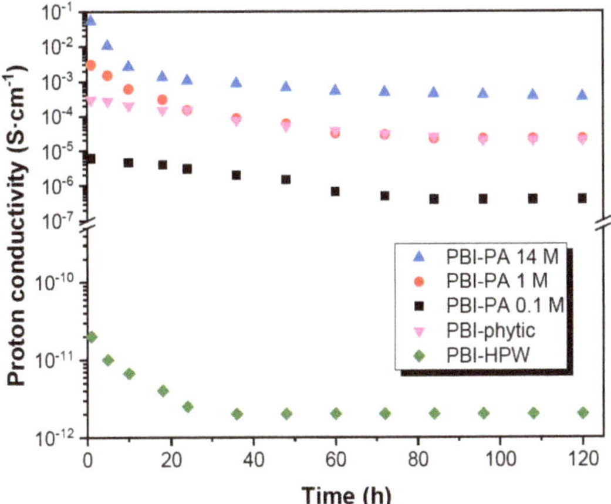

Figure 7. Long-term conductivity study at 160 °C for PBI–PA 0.1 M (■), PBI–PA 1 M (•), PBI–PA 14 M (▲), PBI–phytic acid (0.075 M) (▼), and PBI–HPW (0.1 M) (♦) membranes.

Considering the first cycle, the Bode diagrams in all temperature ranges (Figure S4) showed that the conductivity increased as the temperature increased. As expected, in the PA-doped membranes, the proton conduction depended on the concentration of phosphoric acid molecules. As described, the phosphoric acid molecules can interact with the imidazole ring in the PBI backbone, allowing protons to jump between the PBI network and promote proton transport through hydrogen bond formation and cleavage processes [65,66]. The proton conductivities at 40 °C were 5.6×10^{-7}, 2.9×10^{-4}, and 1.8×10^{-2} S·cm^{-1}, while at 140 °C they were 5.8×10^{-6}, 2.5×10^{-3}, and 5.3×10^{-2} S·cm^{-1} for PBI–PA 0.1 M, PBI–PA 1 M, and PBI–PA 14 M, respectively. It is worth noting that phosphoric acid possesses a high intrinsic proton conductivity, which is mainly attributed to the presence of polarizable hydrogen bonds in a dense network [67]. Conductivities of PBI–phytic acid and PBI–HPW at the same temperatures were 3.2×10^{-5} and 4.1×10^{-6} S·cm^{-1}, which decreased to 2.6×10^{-4} and 1.9×10^{-11} S·cm^{-1}, respectively, at 140 °C. When comparing the conductivity of PBI–phytic acid and PBI–HPW membranes, the phytic acid analogue has a conductivity several orders of magnitude higher than the latter at all temperatures studied. Despite proton conductivities obtained with membranes doped with alternative acids such as phytic acid and HPW being different to those obtained with concentrated phosphoric acid (PBI–PA 14 M or PBI–PA 1 M), some interesting trends can be observed when compared to the sample with a similar concentration of acidic filler (PBI–PA 0.1 M). First, the PBI–HPW membrane displayed higher conductivity than PBI–PA 0.1 M, but lower than PBI–phytic acid membrane at low temperatures (below 80 °C). However, a strong decrease in proton conductivity was observed for the PBI–HPW membrane at 100 °C. A plausible explanation can be attributed to the evaporation of water molecules, which hampers proton transport under anhydrous conditions. Heteropoly acids such as HPW, generally exist in hydrated phases, in which the water molecules can form bridges between ionic clusters to facilitate proton mobility [68]. Secondly, the PBI–phytic acid membrane displayed a higher performance than PBI–PA 0.1 M membranes, both being in the same order of acid concentration. From these results, we can conclude that proton conductivity is several orders of magnitude higher for the membrane of PBI–phytic acid (2.6×10^{-4} S·cm^{-1} at 140 °C) when compared to PBI–HPW (1.9×10^{-11} S·cm^{-1}), and a few orders of magnitude higher than the PBI membrane doped with phosphoric acid at 0.1 M (5.8×10^{-6} S·cm^{-1}). These results may pave the way for the use of this natural acid obtained from plants with unique properties, such as a high phosphate groups content and good chemical and thermal stability [69].

In order to further study the proton conduction mechanism of the membranes, the Arrhenius plots of all the membranes and their proton conduction activation energy values (E_{act}) were analyzed. Figure 8 shows the tendency of the conductivity (σ) with temperature for all the membranes in the range of temperatures between 20 and 200 °C. As observed, proton conductivity increases for all membranes from 20 to 180 °C, following typical Arrhenius behavior. However, for the PBI–HPW membrane, a strong decrease in proton conductivity is observed at 100 °C. Despite HPW having one of the strongest acidities among the different heteropoly acids, its solubility in water is very limited and this drawback has limited its use in PEMFCs, as only a few reports based on PBI [70–72] and other polymeric matrices [73,74] have been described.

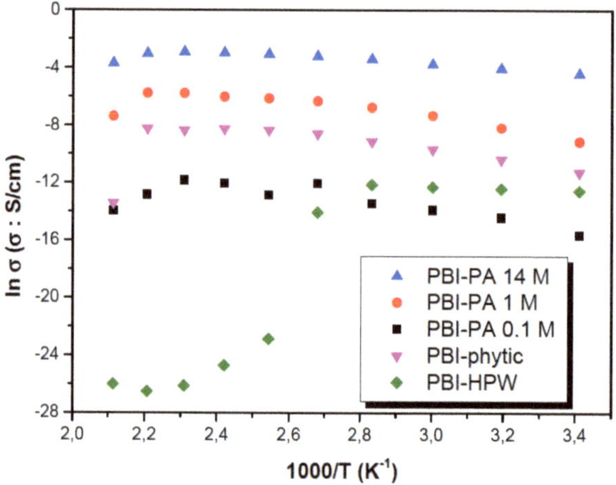

Figure 8. Temperature dependence of proton conductivity (σ) in all the range of temperatures for PBI–PA 0.1 M (■), PBI–PA 1 M (•), PBI–PA 14 M (▲), PBI–phytic acid (0.075 M) (▼), and PBI–HPW (0.1 M) (♦) membranes.

As observed, proton conductivity (σ in $S \cdot cm^{-1}$) followed typical Arrhenius behavior, where a linear tendency can be described in all the range of temperatures following the equation

$$\ln \sigma = \ln \sigma_0 - \frac{E_{act}}{RT} \quad (2)$$

where σ_0 is a pre-exponential factor ($S \cdot cm^{-1}$), E_{act} is the activation energy ($kJ \cdot mol^{-1}$), and R is the ideal gas constant ($8.314\ J \cdot K^{-1} \cdot mol^{-1}$). Using Equation (2), the activation energy was obtained from the slope of the linear fit for each sample. In this regard, the obtained values (see Table 3 for exact values) followed the trend: E_{act} (PBI–PA 14 M) ~12 < E_{act} (PBI–PA 1 M) ≈ E_{act} (PBI–phytic) = ~25 < E_{act} (PBI–PA 0.1 M) = ~28 < < E_{act} (PBI–HPW) = 31 $kJ \cdot mol^{-1}$. These results indicate that proton mobilities increased with the amount of phosphoric acid (i.e., the increase of PA concentration produces a decrease in activation energy which can be attributed to the increasing number of charge density of carriers (protons)). In other words, PA forms channels in the organic phase of porous PBI, as observed by SEM analysis of polymer morphology, facilitating the mobility and consequently, increasing proton conductivity.

Table 3. Proton conductivities (σ) for all membranes under study at 20, 80, and 140 °C (obtained from the first measurement ramp) and calculated activation energies (E_{act}).

Membrane	$\sigma_{20\,°C}$ (S·cm^{-1})	$\sigma_{80\,°C}$ (S·cm^{-1})	$\sigma_{140\,°C}$ (S·cm^{-1})	E_{act} (kJ·mol^{-1})
PBI–DIW	1.1×10^{-14}	1.4×10^{-13}	2.5×10^{-12}	52.6 ± 2.1
PBI–PA 0.1 M	1.6×10^{-6}	2.7×10^{-6}	5.8×10^{-6}	27.5 ± 3.8
PBI–PA 1 M	5.8×10^{-4}	1.2×10^{-3}	2.5×10^{-3}	24.7 ± 3.2
PBI–PA 14 M	2.5×10^{-2}	3.4×10^{-2}	5.3×10^{-2}	11.6 ± 0.7
PBI–phytic acid	1.3×10^{-5}	1.1×10^{-4}	2.6×10^{-4}	25.1 ± 2.3
PBI–HPW	4.8×10^{-6}	2.9×10^{-6}	1.9×10^{-11}	30.8 ± 2.8

Typically, proton conduction in polymeric membranes refers to the process of transport of hydrogen ions in one direction and can be rationalized according to two conduction pathways: the Grotthuss mechanism [75] and the vehicle mechanism [76]. In the Grotthuss mechanism, the proton transport is rationalized by means of the jump of protons in a hydrogen bond network composed by the different groups capable of forming hydrogen bonds, both from the PBI and acidic filler. On the other hand, the proton transport can be explained via the vehicle mechanism through the free phosphoric acid molecules or other acids present in the PBI matrix. The Grotthuss mechanism is characterized by a lower activation energy and according to the calculated activation energy (Table 3), the Grotthuss mechanism dominates the proton transport in acid-doped PBI membranes. Similar activation energies were obtained for PBI–PA 1 M, PBI–PA 0.1 M, and PBI–phytic acid membranes, but PBI–PA 14 M displayed a lower value as the PA concentration is much higher and therefore, the proton transport is more favored. This activation energy is similar to other reported PBI–PA-doped membranes [77,78] and that obtained for an 85% phosphoric acid solution [79].

Assuming that protons are the only available ions that can participate in the charge transport, the diffusivity (D) can be estimated from the conductivity (σ) which was previously determined with the Bode diagrams. According to this consideration, the ionic diffusivity can be estimated applying the Nernst–Einstein equation [80]

$$D = \frac{\sigma RT}{F^2 C_+} \tag{3}$$

where C_+ is the concentration of ions in the membrane, σ the dc-conductivity, R is the gas constant, F is the Faraday constant, and T is the absolute temperature. Considering the stoichiometric amount of acid doping, the calculated density of protons was 5.8×10^{-3}, 19.3×10^{-3}, and 85.1×10^{-3} mol·cm^{-3}, for PBI–PA 0.1 M, PBI–PA 1 M, and PBI–PA 14 M membranes, respectively. On the other hand, the calculated density of protons for the PBI–phytic acid (0.075 M) and PBI–HPW (0.1 M) membranes was 4.2×10^{-3} and 5.2×10^{-3} mol·cm^{-3}, respectively. From these estimated values, the diffusion coefficient of the different membranes can be calculated from Equation (3). The calculated diffusion coefficients (D) at different temperatures are shown in Figure 9. As observed, the diffusion coefficient follows a similar behavior to conductivity from 20 to 180 °C (i.e., typical Arrhenius behavior). Similarly, for PBI–HPW membrane, a strong decrease is observed from 80 °C.

A comparison between the proton diffusion coefficient at 140 °C obtained from Equation (3), shows values of 1.4×10^{-15}, 1.0×10^{-13}, and 1.0×10^{-12} m^2·s^{-1} for PBI–PA 0.1 M, PBI–PA 1 M, and PBI–PA 14 M, respectively. The calculated proton diffusion coefficients (D) increased, as expected, with higher acid concentration. In the case of PBI–phytic acid and PBI–HPW membranes containing acid concentrations of 0.075 and 0.1 M, respectively, diffusion coefficients of 1.4×10^{-14} and 1.0×10^{-21} m^2·s^{-1} were obtained. As observed, for membranes with low acid doping (concentrations around 0.1 M), the proton diffusion coefficient for PBI–phytic acid was higher than that for PBI–PA 0.1 M and PBI–HPW membranes, as observed by its superior proton conduction, also reflected by the differences in the activation energies (Table 3).

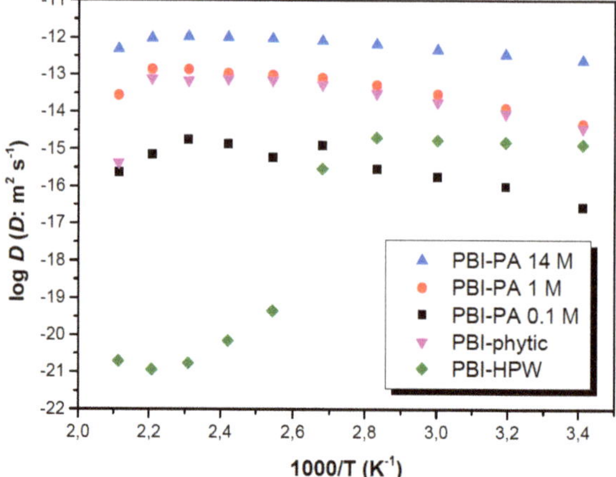

Figure 9. Temperature dependence of the diffusion coefficient (*D*) for PBI–PA 0.1 M (■), PBI–PA 1 M (•), PBI–PA 14 M (▲), PBI–phytic acid (0.075 M) (▼), and PBI–HPW (0.1 M) (♦) membranes.

Finally, a deep look at the variation of diffusion coefficients with temperatures shows that diffusion coefficients increase with temperature in agreement with the increase of proton conductivity. The values found in this work for diffusion coefficients are quite similar to other polymer electrolytes based on polyethylene oxide (PEO) containing $Pr_4N^+I^-$ salts, whose values were around 1.8×10^{-12} $m^2 \cdot s^{-1}$ at temperatures below 100 °C [81]. On the other hand, the values found in this work are 10^4 times higher than the diffusion coefficients of conductive ions obtained in copolymer of vinylidene cyanide and vinyl acetate determined from dielectric measurements using the model of Trukhan at temperatures of 195 °C, where the concentration of electrolytes present in the polymer was around 20×10^{-3} mol cm^{-3} [82]; this is quite similar to our concentrations of 5.8×10^{-3}, 19.3×10^{-3}, 85.1×10^{-3}, 4.2×10^{-3}, and 5.2×10^{-3} mol·cm^{-3} for PBI–PA 0.1 M, PBI–PA 1 M, PBI–PA 14 M, PBI–phytic acid (0.075 M), and PBI–HPW (0.1 M) membranes, respectively.

4. Conclusions

In conclusion, polybenzimidazole membranes containing different acids were prepared and their performance as HT-PEMFCs was evaluated through the analysis of proton conductivity, which was studied by EIS using a blocking electrode configuration. The use of these three acids was reflected in the formation of channels in the polymeric network as observed by cross-section SEM images. These doped materials maintained their mechanical properties and thermal stability for their application as proton exchange membrane fuel cells capable of operating at intermediate or high temperatures. The proton conductivity was strongly dependent on the measurement cycle and decreased along the different cycles. The acid doping increased proton conductivity of PBI membranes reaching values of 0.05 S·cm^{-1} at 140 °C for PBI–PA 14 M. Under low acid doping (concentrations around 0.1 M), membranes doped with phytic acid displayed a superior conducting behavior (2.6×10^{-4} S·cm^{-1}) when compared to doping with phosphoric acid (5.8×10^{-6} S·cm^{-1}) or phosphotungstic acid (1.9×10^{-11} S·cm^{-1}). The results obtained with phytic acid doping may pave the way for the use of this natural acid in combination with other fillers as a sustainable alternative to the use of phosphoric acid and improve its retention in the polymeric membrane. Further applications of this natural acid with PBI membranes are currently under investigation.

Supplementary Materials: The following are available online at http://www.mdpi.com/2073-4360/12/6/1374/s1, Figure S1: FTIR spectra of undoped PBI and PBI doped with PA, phytic acid, and HPW membranes; Figure S2: Proton conductivity of the membranes along the four consecutive measurement cycles; Figure S3: Bode diagrams for the acid-doped membranes for the first ramp of measurement.

Author Contributions: Conceptualization, V.C.; investigation, J.E.; EIS measurements, A.G.-B.; writing—original draft preparation, J.E.; writing—review and editing, J.E., A.G.-B., and V.C.; supervision, V.C.; project administration, V.C.; funding acquisition, V.C. All the authors contributed to the discussions. All authors have read and agreed to the published version of the manuscript.

Funding: This research was funded by the Spanish Ministerio de Economía y Competitividad (MINECO) under the project ENE/2015-69203-R.

Acknowledgments: The authors thanks Santiago V. Luis from Universitat Jaume I for technical assistance with IR measurements, and Enrique Giménez and Óscar Sahuquillo for their technical assistance. Daniel M. Sedgwick from Universitat de València is gratefully acknowledged for the English language revision.

Conflicts of Interest: The authors declare no conflicts of interest.

References

1. Earth's CO2 Home Page. Available online: https://www.co2.earth/ (accessed on 25 March 2020).
2. Kreuer, K.-D. Proton conductivity: Materials and applications. *Chem. Mater.* **1996**, *8*, 610–641. [CrossRef]
3. Steele, B.C.H.; Heinzel, A. Materials for fuel-cell technologies. *Nature* **2001**, *414*, 345–352. [CrossRef]
4. Kreuer, K.D.; Portale, G. A critical revision of the nano-morphology of proton conducting ionomers and polyelectrolytes for fuel cell applications. *Adv. Funct. Mater.* **2013**, *23*, 5390–5397. [CrossRef]
5. Bakangura, E.; Wu, L.; Ge, L.; Yang, Z.; Xu, T. Mixed matrix proton exchange membranes for fuel cells: State of the art and perspectives. *Prog. Polym. Sci.* **2016**, *57*, 103–152. [CrossRef]
6. Konstantinia, D.P.; Fotis, P.; Stylianos, G.N.; Joannis, K.K. Cross-linking of side chain unsaturated aromatic polyethers for high temperature polymer electrolyte membrane fuel cell applications. *Macromolecules* **2011**, *44*, 4942–4951.
7. Armand, M.; Chabagno, J.M.; Duclot, M. Polyethers as solid electrolytes. In *Fast Ion Transport in Solids: Electrodes and Electrolytes*; Vashishta, P., Mundy, J.N., Shenoy, G.K., Eds.; North Holland Publishers: Amsterdan, The Netherlands, 1979.
8. Di Noto, V.; Lavina, S.; Giffin, G.A.; Negro, E.; Scrosati, B. Polymer electrolytes: Present, past and future. *Electrochim. Acta* **2011**, *57*, 4–13. [CrossRef]
9. Tarascon, J.-M.; Armand, M. Issues and challenges facing rechargeable lithium batteries. *Nature* **2001**, *414*, 359–367. [CrossRef]
10. Mauritz, K.A.; Moore, R.B. State of understanding of Nafion. *Chem. Rev.* **2004**, *104*, 4535–4586. [CrossRef]
11. Casciola, M.; Alberti, G.; Sganappa, M.; Narducci, R. On the decay of Nafion proton conductivity at high temperature and relative humidity. *J. Power Sources* **2006**, *162*, 141–145. [CrossRef]
12. Li, Q.; He, R.; Gao, J.-A.; Jensen, J.O.; Bjerrum, N. The CO poisoning effect in PEMFCs operational at temperatures up to 200 °C. *J. Electrochem. Soc.* **2003**, *150*, A1599–A1605. [CrossRef]
13. Jannasch, P. Recent developments in high-temperature proton conducting polymer electrolyte membranes. *Curr. Opin. Colloid Interface Sci.* **2003**, *8*, 96–102. [CrossRef]
14. Purnima, P.; Jayanti, S. Water neutrality and waste heat management in ethanol reformer—HTPEMFC integrated system for on-board hydrogen generation. *Appl. Energy* **2017**, *199*, 169–179. [CrossRef]
15. Hickner, M.A.; Ghassemi, H.; Kim, Y.S.; Einsla, B.R.; McGrath, J.E. Alternative polymer systems for proton exchange membranes (PEMs). *Chem. Rev.* **2004**, *104*, 4587–4612. [CrossRef] [PubMed]
16. Kraytsberg, A.; Ein-Eli, Y. Review of advanced materials for proton exchange membrane fuel cells. *Energy Fuels* **2014**, *28*, 7303–7330. [CrossRef]
17. Reinholdt, M.X.; Kaliaguine, S. Proton exchange membranes for application in fuel cells: Grafted silica/SPEEK nanocomposite elaboration and characterization. *Langmuir* **2010**, *26*, 11184–11195. [CrossRef] [PubMed]
18. Dhanapal, D.; Xiao, M.; Wang, S.; Meng, Y. A review on sulfonated polymer composite/organic-inorganic hybrid membranes to address methanol barrier issue for methanol fuel cells. *Nanomaterials* **2019**, *9*, 668. [CrossRef]

19. Araya, S.S.; Zhou, F.; Liso, V.; Sahlin, S.L.; Vang, J.R.; Thomas, S.; Gao, X.; Jeppesen, C.; Kaer, S.K. A comprehensive review of PBI-based high temperature PEM fuel cells. *Int. J. Hydrog. Energy* **2016**, *41*, 21310–21344. [CrossRef]
20. Asensio, J.A.; Sánchez, E.M.; Gómez-Romero, P. Proton-conducting membranes based on benzimidazole polymers for high-temperature PEM fuel cells. A chemical quest. *Chem. Soc. Rev.* **2010**, *39*, 3210–3239. [CrossRef]
21. Vogel, H.; Marvel, C.S. Polybenzimidazoles: New thermally stable polymers. *J. Polym. Sci.* **1961**, *50*, 511–539. [CrossRef]
22. Wainright, J.S.; Wang, J.-T.; Weng, D.; Savinell, R.F.; Litt, M. Acid-doped polybenzimidazoles: A new polymer electrolyte. *J. Electrochem. Soc.* **1995**, *142*, L121–L123. [CrossRef]
23. Quartarone, E.; Mustarelli, P. Polymer fuel cells based on polybenzimidazole/H3PO4. *Energy Environ. Sci.* **2012**, *5*, 6436–6444. [CrossRef]
24. Wang, L.; Lium, Z.; Ni, J.; Xu, M.; Pan, C.; Wang, D.; Liu, D.; Wang, L. Preparation and investigation of block polybenzimidazole membranes with high battery performance and low phosphoric acid doping for use in high-temperature fuel cells. *J. Membr. Sci.* **2019**, *572*, 350–357. [CrossRef]
25. Wang, L.; Liu, Z.; Liu, Y.; Wang, L. Crosslinked polybenzimidazole containing branching structure with no sacrifice of effective N-H sites: Towards high-performance high-temperature proton exchange membranes for fuel cells. *J. Membr. Sci.* **2019**, *583*, 110–117. [CrossRef]
26. Hu, M.; Li, T.; Neelakandan, S.; Wang, L.; Chen, Y. Cross-linked polybenzimidazoles containing hyperbranched cross-linkers and quaternary ammoniums as high-temperature proton exchange membranes: Enhanced stability and conductivity. *J. Membr. Sci.* **2020**, *593*, 117435. [CrossRef]
27. Ni, J.; Hu, M.; Liu, D.; Xie, H.; Xiang, X.; Wang, L. Synthesis and properties of highly branched polybenzimidazoles as proton exchange membranes for high-temperature fuel cells. *J. Mater. Chem. C* **2016**, *4*, 4814–4821. [CrossRef]
28. Qingfeng, L.; Hjuler, H.A.; Bjerrum, N.J. Phosphoric acid doped polybenzimidazole membranes: Physiochemical characterization and fuel cell applications. *J. Appl. Electrochem.* **2001**, *31*, 773–779. [CrossRef]
29. Samms, S.R.; Wasmus, S.; Savinell, R.F. Thermal stability of proton conducting acid doped polybenzimidazole in simulated fuel cell environments. *J. Electrochem. Soc.* **1996**, *143*, 1225–1232. [CrossRef]
30. Ghosh, S.; Maity, S.; Jana, T. Polybenzimidazole/silica nanocomposites: Organic-inorganic hybrid membranes for PEM fuel cell. *J. Mater. Chem.* **2011**, *21*, 14897–14906. [CrossRef]
31. Escorihuela, J.; García-Bernabé, A.; Montero, A.; Andrio, A.; Sahuquillo, O.; Giménez, E.; Compañ, V. Proton conductivity through polybenzimidazole composite membranes containing silica nanofiber mats. *Polymers* **2019**, *11*, 1182. [CrossRef]
32. Fuentes, I.; Andrio, A.; Garcia-Bernabé, A.; Escorihuela, J.; Viñas, C.; Teixidor, F.; Compañ, V. Structural and dielectric properties of cobaltacarborane composite polybenzimidazole membranes as solid polymer electrolytes at high temperature. *Phys. Chem. Chem. Phys.* **2018**, *20*, 10173–10184. [CrossRef]
33. Ozdemir, Y.; Uregen, N.; Devrim, Y. Polybenzimidazole based nanocomposite membranes with enhanced proton conductivity for high temperature PEM fuel cells. *Int. J. Hydrog. Energy* **2017**, *42*, 2648–2657. [CrossRef]
34. Uregen, N.; Pehlivanoglu, K.; Ozdemir, Y.; Devrim, Y. Development of polybenzimidazole/graphene oxide composite membranes for high temperature PEM fuel cells. *Int. J. Hydrog. Energy* **2017**, *42*, 2636–2647. [CrossRef]
35. Reyes-Rodriguez, J.L.; Escorihuela, J.; Garcia-Bernabé, A.; Giménez, E.; Solorza-Feria, O.; Compañ, V. Proton conducting electrospun sulfonated polyether ether ketone graphene oxide composite membranes. *RSC Adv.* **2017**, *7*, 53481–53491. [CrossRef]
36. Escorihuela, J.; Sahuquillo, O.; García-Bernabé, A.; Giménez, E.; Compañ, V. Phosphoric acid doped polybenzimidazole (PBI)/zeolitic imidazolate framework composite membranes with significantly enhanced proton conductivity under low humidity conditions. *Nanomaterials* **2018**, *8*, 775. [CrossRef] [PubMed]
37. Barjola, A.; Escorihuela, J.; Andrio, A.; Giménez, E.; Compañ, V. Enhanced conductivity of composite membranes based on sulfonated poly(ether ether ketone) (SPEEK) with zeolitic imidazolate frameworks (ZIFs). *Nanomaterials* **2018**, *8*, 1042. [CrossRef]

38. Escorihuela, J.; Narducci, R.; Compañ, V.; Costantino, F. Proton conductivity of composite polyelectrolyte membranes with metal-organic frameworks for fuel cell applications. *Adv. Mater. Interfaces* **2019**, *6*, 1801146. [CrossRef]
39. Liu, S.; Zhou, L.; Wang, P.; Zhang, F.; Yu, S.; Shao, Z.; Yi, B. Ionic-liquid-based proton conducting membranes for anhydrous H2/Cl2 fuel-cell applications. *ACS Appl. Mater. Interfaces* **2014**, *6*, 3195–3200. [CrossRef]
40. Kallem, P.; Eguizabal, A.; Mallada, R.; Pina, M.P. Constructing straight polyionic liquid microchannels for fast anhydrous proton transport. *ACS Appl. Mater. Interfaces* **2016**, *8*, 35377–35389. [CrossRef]
41. Escorihuela, J.; García-Bernabé, A.; Montero, A.; Sahuquillo, O.; Giménez, E.; Compañ, V. Ionic liquid composite polybenzimidazol membranes for high temperature PEMFC Applications. *Polymers* **2019**, *11*, 732. [CrossRef]
42. Mack, F.; Aniol, K.; Ellwein, C.; Kerres, J.; Zeis, R. Novel phosphoric acid-doped PBI-blends as membranes for high-temperature PEM fuel cells. *J. Mater. Chem. A* **2015**, *3*, 10864–10874. [CrossRef]
43. Li, Z.; He, G.; Zhang, B.; Cao, Y.; Wu, H.; Jiang, Z.; Tiantian, Z. Enhanced proton conductivity of Nafion hybrid membrane under different humidities by incorporating metal–organic frameworks with high phytic acid loading. *ACS Appl. Mater. Interfaces* **2014**, *6*, 9799–9807. [CrossRef] [PubMed]
44. Tanaka, M.; Takeda, Y.; Wakiya, T.; Wakamoto, Y.; Harigaya, K.; Ito, T.; Tarao, T.; Kawakami, H. Acid-doped polymer nanofiber framework: Three-dimensional proton conductive network for high-performance fuel cells. *J. Power Sources* **2017**, *342*, 125–134. [CrossRef]
45. Zeng, J.; Zhou, Y.; Li, L.; Jiang, S.P. Phosphotungstic acid functionalized silica nanocomposites with tunable bicontinuous mesoporous structure and superior proton conductivity and stability for fuel cells. *Phys. Chem. Chem. Phys.* **2011**, *13*, 10249–10257. [CrossRef]
46. Zhou, Y.; Yang, J.; Su, H.; Zeng, J.; Jiang, S.P.; Goddard, W.A. Insight into proton transfer in phosphotungstic acid functionalized mesoporous silica-based proton exchange membrane fuel cells. *J. Am. Chem. Soc.* **2014**, *136*, 4954–4964. [CrossRef] [PubMed]
47. Liu, X.; Li, Y.; Xue, J.; Zhu, W.; Zhang, J.; Yin, Y.; Qin, Y.; Jiao, K.; Du, Q.; Cheng, B.; et al. Magnetic field alignment of stable proton-conducting channels in an electrolyte membrane. *Nat. Commun.* **2019**, *10*, 842. [CrossRef]
48. Zhai, L.; Li, H. Polyoxometalate–Polymer Hybrid Materials as Proton Exchange Membranes for Fuel Cell Applications. *Molecules* **2019**, *24*, 3425. [CrossRef]
49. Yuan, J.; Zhou, G.; Pu, H. Preparation and properties of Nafion®/hollow silica spheres composite membranes. *J. Membr. Sci.* **2008**, *325*, 742–748. [CrossRef]
50. Zhang, X.; Fu, X.; Yang, S.; Zhang, Y.; Zhang, R.; Hu, S.; Bao, X.; Zhao, F.; Li, X.; Liu, Q. Design of sepiolite-supported ionogel-embedded composite membranes without proton carrier wastage for wide-temperature-range operation of proton exchange membrane fuel cells. *J. Mater. Chem. A* **2019**, *7*, 15288–15301. [CrossRef]
51. Wang, S.; Zhao, C.J.; Ma, W.J.; Zhang, G.; Liu, Z.G.; Ni, J.; Li, M.Y.; Zhang, N.; Na, H. Preparation and properties of epoxy-cross-linked porous polybenzimidazole for high temperature proton exchange membrane fuel cells. *J. Membr. Sci.* **2012**, *411–412*, 54–63. [CrossRef]
52. Bose, S.; Kuila, T.; Nguyen, T.X.H.; Kim, N.H.; Lau, K.; Lee, J.H.P. Polymer membranes for high temperature proton exchange membrane fuel cell: Recent advances and challenges. *Prog. Polym. Sci.* **2011**, *36*, 813–843. [CrossRef]
53. Choi, S.-W.; Park, J.O.; Pak, C.; Choi, K.H.; Lee, J.-C.; Chang, H. Design and synthesis of cross-linked copolymer membranes based on poly(benzoxazine) and polybenzimidazole and their application to an electrolyte membrane for a high-temperature PEM fuel cell. *Polymers* **2013**, *5*, 77–111. [CrossRef]
54. Lua, Z.; Lugo, M.; Santare, M.H.; Karlsson, A.M.; Busby, F.C.; Walsh, P. An experimental investigation of strain rate, temperature and humidity effects on the mechanical behavior of a perfluorosulfonic acid membrane. *J. Power Sources* **2012**, *214*, 130–136. [CrossRef]
55. Yang, J.S.; Li, Q.F.; Cleemann, L.N.; Xu, C.X.; Jensen, J.O.; Pan, C.; Bjerrumb, N.J.; He, R.H. Synthesis and properties of poly(aryl sulfone benzimidazole) and its copolymers for high temperature membrane electrolytes for fuel cells. *J. Mater. Chem.* **2012**, *22*, 11185–11195. [CrossRef]
56. Sana, B.; Unnikrishnan, G.; Jana, T.; KS, S.K. Polybenzimidazole co-polymers: Their synthesis, morphology and high temperature fuel cell membrane properties. *Polym. Chem.* **2020**, *11*, 1043–1054.

57. Li, Q.; Pan, C.; Jensen, J.O.; Noyé, P.; Bjerrum, N.J. Cross-Linked Polybenzimidazole Membranes for Fuel Cells. *Chem. Mater.* **2007**, *19*, 350–352. [CrossRef]
58. Mader, J.A.; Benicewicz, B.C. Synthesis and properties of segmented block copolymers of functionalised polybenzimidazoles for high-temperature PEM fuel cells. *Fuel Cells* **2011**, *11*, 222–237. [CrossRef]
59. Gao, C.; Hu, M.; Wang, L.; Wang, L. Synthesis and properties of phosphoric-acid-doped polybenzimidazole with hyperbranched cross-linkers decorated with imidazolium groups as high-temperature proton exchange membranes. *Polymers* **2020**, *12*, 515. [CrossRef]
60. Xiao, L.; Apple, T.; Zhang, H.; Scanlon, E.; Ramanathan, L.S.; Choe, E.-W.; Rogers, D.; Benicewicz, B.C. High-temperature polybenzimidazole fuel cell membranes via a sol-gel process. *Chem. Mater.* **2005**, *17*, 5328–5333. [CrossRef]
61. Sacco, A. Electrochemical impedance spectroscopy: Fundamentals and application in dye-sensitized solar cells. *Renew. Sust. Energy Rev.* **2017**, *79*, 814–829. [CrossRef]
62. Randviir, E.P.; Banks, C.E. Electrochemical impedance spectroscopy: An overview of bioanalytical applications. *Anal. Methods* **2013**, *5*, 1098–1115. [CrossRef]
63. Gomadam, P.M.; Weidner, J.W. Analysis of electrochemical impedance spectroscopy in proton exchange membrane fuel cells. *Int. J. Energy Res.* **2005**, *29*, 1133–1151. [CrossRef]
64. Donne, S.W. General principles of electrochemistry. In *Supercapacitors: Materials, Systems, and Applications;* Béguin, F., Frąckowiak, E., Eds.; Wiley-VCH Verlag GmbH & Co. KGaA: Weinheim, Germany, 2013; pp. 1–68.
65. Nawn, G.; Pace, G.; Lavina, S.; Vezzù, K.; Negro, E.; Bertasi, F.; Polizzi, S.; Di Noto, V. Interplay between composition, structure, and properties of new H_3PO_4-doped PBI_4N-HfO_2 nanocomposite membranes for high-temperature proton exchange membrane fuel cells. *Macromolecules* **2015**, *48*, 15–27. [CrossRef]
66. Liu, F.; Wang, S.; Chen, H.; Li, J.; Tian, X.; Wang, X.; Mao, T.; Xu, J.; Wang, Z. Cross-linkable polymeric ionic liquid improve phosphoric acid retention and long-term conductivity stability in polybenzimidazole based PEMs. *ACS Sustain. Chem. Eng.* **2018**, *6*, 16352–16362. [CrossRef]
67. Vilčiauskas, L.; Tuckerman, M.E.; Bester, G.; Paddison, S.J.; Kreuer, K.-D. The Mechanism of Proton Conduction in Phosphoric Acid. *Nat. Chem.* **2012**, *4*, 461–466. [CrossRef]
68. Bose, A.B.; Gopu, S.; Li, W. Enhancement of proton exchange membrane fuel cells performance at elevated temperatures and lower humidities by incorporating immobilized phosphotungstic acid in electrodes. *J. Power Sources* **2014**, *263*, 217–222. [CrossRef]
69. Crea, F.; De Stefano, C.; Milea, D.; Sammartano, S. Formation and stability of phytate complexes in solution. *Coord. Chem. Rev.* **2008**, *252*, 1108–1120. [CrossRef]
70. Lua, J.L.; Fang, Q.H.; Li, S.L.; Jiang, S.P. A novel phosphotungstic acid impregnated meso-Nafion multilayer membrane for proton exchange membrane fuel cells. *J. Membr. Sci.* **2013**, *427*, 101–107. [CrossRef]
71. Hasani-Sadrabadi, M.M.; Dashtimoghadam, E.; Majedi, F.S.; Moaddel, H.; Bertsch, A.; Renaudand, P. Superacid-doped polybenzimidazole-decorated carbon nanotubes: A novel high-performance proton exchange nanocomposite membrane. *Nanoscale* **2013**, *5*, 11710–11717. [CrossRef]
72. Wang, S.; Sun, P.; Li, Z.; Liu, G.; Yin, X. Comprehensive performance enhancement of polybenzimidazole based high temperature proton exchange membranes by doping with a novel intercalated proton conductor. *Int. J. Hydrog. Energy* **2018**, *43*, 9994–10003. [CrossRef]
73. Kim, A.R.; Vinothkannan, M.; Kim, J.S.; Yoo, D.J. Proton-conducting phosphotungstic acid/sulfonated fluorinated block copolymer composite membrane for polymer electrolyte fuel cells with reduced hydrogen permeability. *Polym. Bull.* **2018**, *75*, 2779–2804. [CrossRef]
74. Kim, A.R.; Park, C.J.; Vinothkannan, M.; Yoo, D.J. Sulfonated poly ether sulfone/heteropoly acid composite membranes as electrolytes for the improved power generation of proton exchange membrane fuel cells. *Compos. Part B Eng.* **2018**, *155*, 272–281. [CrossRef]
75. Agmon, N. The Grotthuss mechanism. *Chem. Phys. Lett.* **1995**, *244*, 456–462. [CrossRef]
76. Kreuer, K.D.; Rabenau, A.; Weppner, W. Vehicle mechanism, a new model for the interpretation of the conductivity of fast proton conductors. *Angew. Chem. Int. Ed. Engl.* **1982**, *21*, 208–209. [CrossRef]
77. Xu, C.; Cao, Y.; Kumar, R.; Wu, X.; Wang, X.; Scott, K. A polybenzimidazole/sulfonated graphite oxide composite membrane for high temperature polymer electrolyte membrane fuel cells. *J. Mater. Chem.* **2011**, *21*, 11359–11364. [CrossRef]

78. Rewar, A.S.; Chaudhari, H.D.; Illathvalappil, R.; Sreekumar, K.; Kharul, U.K. New approach of blending polymeric ionic liquid with polybenzimidazole (PBI) for enhancing physical and electrochemical properties. *J. Mater. Chem. A* **2014**, *2*, 14449–14458. [CrossRef]
79. He, R.; Li, Q.; Xiao, G.; Bjerrum, N.J. Proton conductivity of phosphoric acid doped polybenzimidazole and its composites with inorganic proton conductors. *J. Membr. Sci.* **2003**, *226*, 169–184. [CrossRef]
80. Monroe, C.W. Ionic Mobility and Diffusivity. In *Encyclopedia of Applied Electrochemistry*; Kreysa, G., Ota, K., Savinell, R.F., Eds.; Springer: New York, NY, USA, 2014.
81. Bandara, T.M.W.J.; Dissanayake, M.A.K.L.; Alboinsson, I.; Mellander, B.-E. Mobile charge carrier concentration and mobility of a polymer electrolyte containing PEO and $Pr_4N^+I^-$ using electrical and dielectric measurements. *Solid State Ion.* **2011**, *189*, 63–68. [CrossRef]
82. Compañ, V.; Sorensen, T.S.; Diaz-Calleja, R.; Riande, E. Diffusion coefficients of conductive ions in a copolymer of vinylidene cyanide and vinyl acetate obtained from dielectric measurements using the model of Trukhan. *J. Appl. Phys.* **1996**, *79*, 403–411. [CrossRef]

Article

Crosslinked Sulfonated Polyphenylsulfone-Vinylon (CSPPSU-vinylon) Membranes for PEM Fuel Cells from SPPSU and Polyvinyl Alcohol (PVA)

Je-Deok Kim [1,2,3,*], Satoshi Matsushita [1] and Kenji Tamura [3]

1 Polymer Electrolyte Fuel Cell Group, Global Research Center for Environmental and Energy Based on Nanomaterials Science (GREEN),Tsukuba Ibaraki 305-0044, Japan; satoshi.matsushita@agc.com

2 Hydrogen Production Materials Group, Center for Green Research on Energy and Environmental Materials, Tsukuba Ibaraki 305-0044, Japan

3 Functional Clay Materials Group, Research Center for Functional Materials, National Institute for Materials Science (NIMS), 1-1 Namiki, Tsukuba, Ibaraki 305-0044, Japan; Tamura.Kenji@nims.go.jp

* Correspondence: kim.jedeok@nims.go.jp; Tel.: +81-29-860-4764; Fax: +81-29-860-4984

Received: 3 June 2020; Accepted: 14 June 2020; Published: 16 June 2020

Abstract: A crosslinked sulfonated polyphenylsulfone (CSPPSU) polymer and polyvinyl alcohol (PVA) were thermally crosslinked; then, a CSPPSU-vinylon membrane was synthesized using a formalization reaction. Its use as an electrolyte membrane for fuel cells was investigated. PVA was synthesized from polyvinyl acetate (PVAc), using a saponification reaction. The CSPPSU-vinylon membrane was synthesized by the addition of PVA (5 wt%, 10 wt%, 20 wt%), and its chemical, mechanical, conductivity, and fuel cell properties were studied. The conductivity of the CSPPSU-10vinylon membrane is higher than that of the CSPPSU membrane, and a conductivity of 66 mS/cm was obtained at 120 °C and 90% RH (relative humidity). From a fuel cell evaluation at 80 °C, the CSPPSU-10vinylon membrane has a higher current density than CSPPSU and Nafion212 membranes, in both high (100% RH) and low humidification (60% RH). By using a CSPPSU-vinylon membrane instead of a CSPPSU membrane, the conductivity and fuel cell performance improved.

Keywords: PPSU; SPPSU; PVA; CSPPSU-vinylon; PEMFCs

1. Introduction

In order to realize a low-carbon society that includes highly efficient energy systems which make effective use of renewable energy, economically sustainable growth, environmental protection, and energy security are required. Energy conversion/storage devices—such as fuel cells, water electrolysis, secondary batteries, and solar cells—are core technologies for building a low-carbon society. In order to produce these devices safely and with high performances, their constituent materials must have high performances. Polymer electrolyte membranes for proton exchange in fuel cells are required for polymer electrolyte membrane fuel cells (PEMFCs), and polymer electrolyte membrane water electrolysis (PEMWE), and Nafion membranes mainly using perfluorosulfonic acid (PFSA) ion exchange resins, are often used. Nafion membranes have high proton conductivities and excellent chemical stabilities, and have been used in both mobile and stationary fuel cells [1,2]. However, the performance of these fuel cells suffers from the deterioration of their mechanical properties, due to thinning of the electrolyte membrane. In addition, higher operating temperatures are required to improve both the glass transition temperature (T_g) and the proton conductivity at high temperatures. At the same time, research on hydrocarbon-based electrolytes, instead of fluorine-based electrolytes, has been conducted. Non-fluorine-based electrolytes are low cost materials, have high T_g values, and have been studied for many years. However, their performances and chemical stabilities, which are lower than those of fluorine-based ones, have been the biggest obstacles towards their practical use. Thus, materials for proton exchange electrolytes with higher performances are still needed.

The drawbacks of PFSA membranes have prompted research into alternative membranes. Various aromatic polymer ionomer membranes are being actively investigated. Sulfonated polyphenylsulfone (SPPSU) [3–20], sulfonated polyetheretherketone (SPEEK) [21–32], sulfonated polysulfone (SPSU) [33–38], sulfonated polyphenylene sulfone (SPPS) [39], sulfonated polyphenylene (SPP) [40], sulfonated polyethersulfone (SPES) [41,42], sulfonated polyimide (SPI) [43–45], sulfonated polyphenylene oxide (SPPO) [46], and polybenzimidazole (PBI) [47,48] are attracting special interest.

We are developing a crosslinked sulfonated polyphenylsulfone (CSPPSU) membrane using the sulfonation of the polyphenylsulfone (PPSU) polymer, which has an excellent thermal stability, high chemical resistance, and is low cost [3,5,6,19,20]. According to a fuel cell evaluation using the CSPPSU membrane, the membrane could be used for 4000 h [19]. However, compared to fluorine-based electrolyte membranes, hydrocarbon-based CSPPSU membranes are still insufficient in terms of high performances and high durabilities in fuel cells. It is necessary to further improve the performance and durability of SPPSU membranes. In this study, we prepared a CSPPSU-vinylon membrane, which has a higher performance than both CSPPSU and Nafion212 membranes. Vinylon was obtained using a formalization reaction with polyvinyl alcohol (PVA), and PVA was obtained by the saponification of polyvinyl acetate (PVAc). Crosslinked SPPSU-vinylon membranes from the SPPSU and PVA, were obtained using thermal crosslinking and a formalization reaction. The vinylon was somewhat stable, even in a high-hydration environment. First, we prepared a crosslinked SPPSU-vinylon membrane using thermal crosslinking and the vinylonization of a SPPSU-PVA composite, and these methods appeared promising for the thinning of other polymer electrolyte membranes.

2. Experimental

2.1. Materials

Polyvinyl acetate (PVAc, $(C_4H_6O_2)_n$, Mw = 100,000) was purchased from Sigma-Aldrich Corporation (St. Louis, MO, USA). A DuPontTM Nafion212 membrane (NR-212) was purchased from DuPont (USA). Methanol (CH_3OH), formaldehyde (CH_2O, 37%), sodium chloride (NaCl), sodium hydroxide (NaOH), and sulfuric acid (H_2SO_4) were purchased from Nacalai Tesque, Inc. Polyphenylsulfone (Solvay Radel R-5000 NT) (Mn = 26,000; Mw = 50,000; Mw/Mn = 1.9) was provided by Solvay Specialty Polymers Japan K.K. (glass transition temperature (T_g) = 220 °C, (Tokyo, Japan). A dialysis tubing cellulose membrane, which has a molecular weight cut-off (MWCO) of 14,000, and dimethyl sulfoxide (DMSO) were purchased from Sigma-Aldrich Co., Ltd. Deionized (DI) water was obtained using a PURELAB® Option-R 7 ELGA LabWater, at 15 Mohm cm and 25 °C. Sodium sulfate (Na_2SO_4) and Iron (II) chloride tetrahydrate ($FeCl_2{\cdot}4H_2O$) were purchased from Fujifilm Wako Pure Chemical Corporation (Osaka, Japan).

2.2. Synthesis of SPPSU and PVA, and Preparation of CSPPSU-vinylon Membranes

The synthesis and properties of SPPSU (Mw ≈ 150,000) have been described in detail in previous reports [3,19]. PVA was synthesized using the following method. PVAc (1 g) was dissolved in a flask with methanol (50 mL). Then, a 40% NaOH solution was added, and the mixture was allowed to react at 40 °C for 10 min (saponification). The reaction mixture was washed with methanol 4 times and filtered. To remove the remaining solvent, the product was dried at 80 °C for 24 h, and PVA (0.5 g, Mw ≈ 50,000) was obtained. Crosslinked SPPSU-vinylon membranes were obtained from SPPSU and PVA using the following method. A glass vial was charged with SPPSU (0.5 g) and DMSO (20 mL), and dissolved. PVA (0.025 g, 0.05 g, 0.1 g) and DMSO (4 mL) were put into another glass vial and dissolved. Then, the PVA-DMSO solution was added to the SPPSU-DMSO solution, and the mixture was stirred for 1 h. The SPPSU-PVA-DMSO solution was transferred to a glass container, dried for 24 h at 80 °C, and then annealed in air at 120 °C (24 h), 160 °C (24 h), and 180 °C (24 h). Next, the vinylon from the PVA was prepared using a formalization solution (H_2O:H_2SO_4:Na_2SO_4:CH_2O = 1.00:0.21:0.20:0.06 in mass ratio) reaction, for 2 h at 60 °C. Activation was performed using the following procedure: heating in 0.5 M NaOH at 80 °C overnight, washing in DI H_2O, heating at 1M H_2SO_4 at 80 °C for 2 h, and washing in DI H_2O for 2 h. Finally, the crosslinked SPPSU-vinylon membranes were dried at room temperature before use. The crosslinked SPPSU-vinylon membranes were very flexible and dark brown. The classification of the crosslinked SPPSU-vinylon membranes is shown in Table 1.

Table 1. Classification of crosslinked sulfonated polyphenylsulfone-vinylon (CSPPSU-vinylon) membranes.

Varied Parameter	Variable Parameter	Membrane Classification
PVA loading (wt%)	0	CSPPSU
	5	CSPPSU-5vinylon
	10	CSPPSU-10vinylon
	20	CSPPSU-20vinylon

2.3. *Iron-Exchange Chromotography (IEC), D.S. (Degree of Sulfonation), Water-Uptake (W.U.), λ, and Crosslink Rates ($D_{crosslink}$)*

The IEC values were determined using the following equation: IEC (meq/g) = cv/W_{dry}, where c (mmol/L) is the concentration of standardized NaOH aq. used for titration (0.01mol/L), v (L) is the volume of standardized NaOH aq. used for titration, and W_{dry} (g) is the mass of the dry membrane. The water-uptake (W.U.) of the membranes at room temperature was calculated using the following: W.U. (%) = $[(W_{wet} - W_{dry})/W_{dry}] \times 100$, where W_{wet} is the mass of the wet membrane. The hydration number (λ) for the membranes was determined using the following: λ ($[H_2O]/[SO_3H]$) = $[1000(W_{wet} - W_{dry})]/18W_{dry}IEC$. The degree of crosslinking (crosslink rate, $D_{crosslink}$) in the membranes was determined using the following: $D_{crosslink}$ (%) = $[(IEC_{before\ annealing} - IEC_{after\ annealing})/IEC_{before\ annealing}] \times 100$ [19].

2.4. *Oxidative Stability (Fenton's Test)*

The oxidative stabilities of the membranes were evaluated by immersing a small piece of the membrane into Fenton's reagent [3 wt% H_2O_2 and 2 ppm Fe(II) (added as $FeCl_2 \cdot 4H_2O$)], at ~80 °C for 1 h while stirring. The samples were dried at 80 °C before the measurements. The membranes were repeatedly washed with DI H_2O, and dried at 80 °C overnight following the reaction. The oxidative stabilities were determined as follows: [(mass of residual membrane after the test)/(initial mass of membrane)] × 100.

2.5. *Chemical Structure of the Samples*

Fourier-transform infrared (FTIR) absorption spectra of the samples were obtained on a Thermo Scientific Nicolet 6700 spectrometer, in an attenuated total reflection (ATR) mode.

2.6. *Mechanical and Thermal Behavior of the Samples*

Stress-strain tests on the membranes were accomplished using a Tension Test Machine (Shimazu, EZ-S) at room temperature [19]. The thermal and mass properties of the membranes were investigated using thermogravimetric and mass analyses with a Thermoplus TG8120 TG-DTA/H (Rigaku Co. Ltd., Japan). The samples were heated from 60 to 800 °C at 5 °C/min in air, after keeping them for 1 h at 60 °C.

2.7. Conductivity and Single Cell Measurements of the Samples

The proton conductivities of the membranes were evaluated using a four-point probe impedance spectroscopy. For the membrane electrode assembly (MEA), the thickness of the membranes was approximately 50 μm, and a Pt/C/ionomer (ionomer/carbon = 1) catalyst electrode (EIWA corporation) containing 0.3 mg/cm^2 of Pt on a GDL electrode (Sigracet® GDL 25BC of SGL Group Co. Ltd., Japan), was used. The effective electrode area of the single cell was 4 cm^2. The MEA was acquired by loading a membrane between the anode and cathode, and hot-pressing at 130 °C and ~9.8 kN for 20 min. A single cell performance was measured in relation to the amount of hydrogen (H_2) and oxygen (O_2) at the anode and cathode, respectively, at 80 °C, 100% and 60% RH (relative humidity), and ambient pressure. The gas flow rate of hydrogen and oxygen were 50 cc/min and 100 cc/min, respectively. Linear sweep voltammetry (LSV) was evaluated in the potential range of 0.02–0.5 V at 2 mV/s [19].

3. Results and Discussion

3.1. CSPPSU-vinylon Membranes

Thermally crosslinked membranes of SPPSU polymers have been reported in previous research [5,19]. Crosslinking occurs between the sulfone groups of SPPSU under a thermal environment. The same phenomenon occurs in composite membranes of SPPSU polymers and PVA polymers. In addition, crosslinking occurs between the sulfone groups of SPPSU and the hydroxy groups of PVA upon heat treatment. A crosslinked SPPSU-vinylon membrane could be obtained using a formalization reaction with PVA (Figure 1).

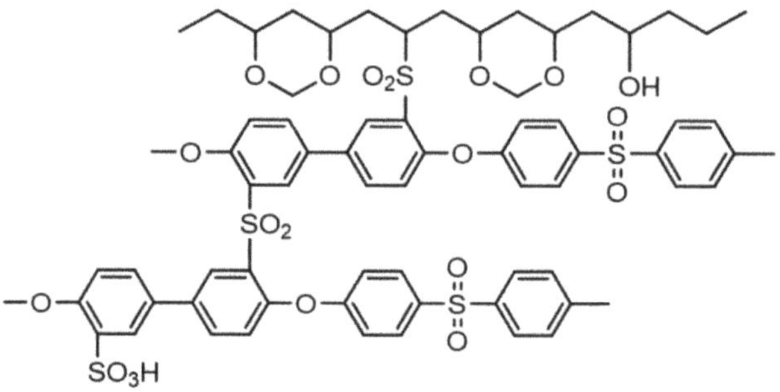

Figure 1. Schematic diagram of CSPPSU-vinylon membrane.

PVA was synthesized by hydrolyzing PVAc (saponification). The peak for the carbonyl groups (C=O) of the PVAc appeared at 1728 cm^{-1} in the IR spectrum (Figure 2a). For PVA synthesized using the hydrolysis of PVAc, the peak due to the carbonyl group disappeared, and new peaks for the OH groups appeared at 3272 cm^{-1}, 1655 cm^{-1}, and 1324 cm^{-1} (Figure 2b). These results indicate that PVA can be obtained by hydrolyzing PVAc. The IR spectra of the CSPPSU-vinylon membranes did not change significantly with the amount of PVA added and had similar characteristics (Figure 2c–e). The peaks for both SPPSU and PVA appeared in the spectra. The crosslinking of SPPSU has been reported in more detail in a previous paper [3,5]. In the IR spectra, it was not clear whether the sulfone bridge of SPPSU and the hydroxy group of PVA were crosslinked using hydrolysis to form a sulfone bridge (-SO_2-). Moreover, it was difficult to determine whether PVA had been changed to vinylon. However, we can assume that crosslinking and vinylon formation progressed, as the appearance of the obtained membranes were very uniform and flexible. Figure 2 and Table 2 show the FTIR spectra and summarize the assignments of the peaks, respectively.

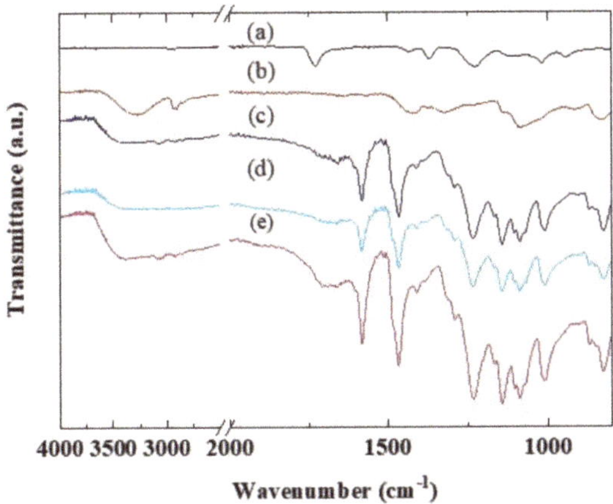

Figure 2. Fourier transform infrared (FTIR) properties of (**a**) polyvinyl acetate (PVAc), (**b**) syn. polyvinyl alcohol (PVA), (**c**) CSPPSU-5vinylon, (**d**) CSPPSU-10vinylon, and (**e**) CSPPSU-20vinylon membranes with different amounts of PVA.

Table 2. Summary of assignments of the FTIR spectra of PVAc, syn. PVA, and CSPPSU-vinylon membranes.

Polymer	PVAc	PVA	CSPPSU-vinylon
υH-O-H (cm^{-1})		3272	3421
Aromatic υC-H (cm^{-1})			3093, 3074, 3038
Aliphatic υC-H (cm^{-1})	2974, 2928, 2852, 1431, 1367	2932, 2898, 1424	2920, 2846
δs, H-O-H (cm^{-1})		1655	1714, 1665
Aromatic υC=C (cm^{-1})			1584, 1469
υC=O (cm^{-1})	1728		
δs, C-O-H (cm^{-1})		1324	1324
υas, C-O-C (cm^{-1})			1232
Aliphatic υC-O (cm^{-1})	1220, 1116, 1016, 939	1568, 1083	
υs, O=S=O (cm^{-1})			1142

3.2. Thermal and Mechanical Properties of the CSPPSU-vinylon Membranes

The thermal (Figure 3) and mechanical properties (Figure 4) of the CSPPSU-vinylon membranes, prepared by varying the amount of PVA added, were investigated. The CSPPSU-vinylon membrane exhibited a lower desorption of the sulfone groups and had a lower decomposition temperature of the polymer backbone than the CSPPSU membrane (Figure 3, Table 3). The thermal behavior of the CSPPSU-vinylon membrane on the amount of PVA added was similar. As for the weight reduction ratio of water due to water vaporization, the CSPPSU-vinylon membrane had a higher water content than the CSPPSU membrane (Table 3). In the TGA (thermal gravimetric analysis) curves for the CSPPSU sample, residuals (inorganic substances) appeared after 600 °C. However, every CSPPSU-vinylon sample burned at 600 °C. This suggests that the SPPSU and the vinylon were crosslinked into one polymer.

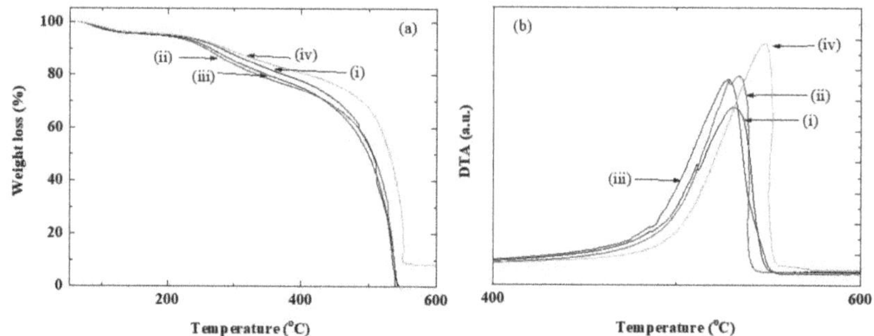

Figure 3. (**a**) TG (thermal gravimetric) and (**b**) DTA (differential thermal analysis) results of (i) CSPPSU-5vinylon, (ii) CSPPSU-10vinylon, (iii) CSPPSU-20vinylon, and (iv) CSPPSU membranes.

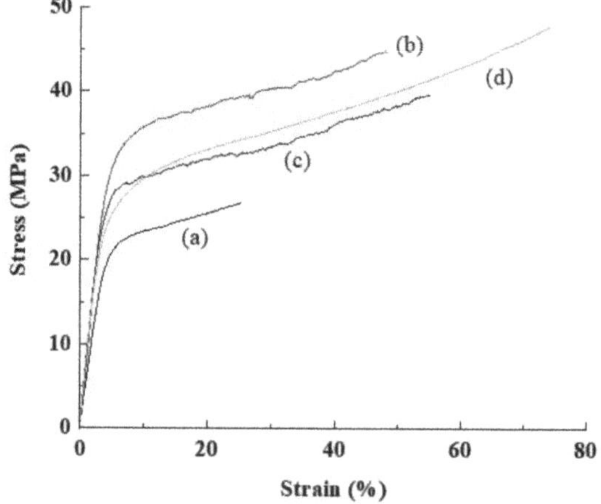

Figure 4. Stress-strain properties of (**a**) CSPPSU-5vinylon, (**b**) CSPPSU-10vinylon, (**c**) CSPPSU-20vinylon, and (**d**) CSPPSU membranes.

Table 3. Summary of temperature ranges and mass losses observed for each step in the TG-DTA curves for CSPPSU-vinylon membranes.

Sample Name	Evaporation of H_2O Interacting with $-SO_3H$ or -OH Group		Desubstitution of $-SO_3H$ Group		Thermal Decomposition of Polymer Backbone
	ΔT (°C)	ΔWt. Loss (%)	ΔT (°C)	ΔWt. Loss (%)	Peak of Exothermic (°C)
CSPPSU	61–210	4.5	210–453	19.1	548
CSPPSU-5vinylon	61–212	4.9	212–403	17.5	532
CSPPSU-10vinylon	61–197	5.0	197–389	19.5	534
CSPPSU-20vinylon	61–197	5.0	197–400	19.8	528

On the other hand, the dependence of the CSPPSU-vinylon membrane on the amount of added PVA, was noticeable in the evaluation of its mechanical properties (Figure 4, Table 4). The CSPPSU-10vinylon membranes obtained by adding 10 wt% PVA to SPPSU had higher tensile strengths than the other crosslinked membranes. However, the tensile elongation increased with an increase in the amount of added PVA, and the tensile strength and tensile elongation of the CSPPSU-5vinylon membrane containing 5 wt% PVA, was low in comparison to the other membranes. The flexural modulus of the membrane decreased with an increase in the amount of PVA. The tensile elongation characteristics of the CSPPSU-vinylon membrane were smaller than those of the CSPPSU membranes. The favorable tensile elongation of the CSPPSU-10vinylon membrane may be due to its better homogeneity in comparison to the other membranes.

Table 4. Mechanical properties of CSPPSU-vinylon membranes.

	CSPPSU	CSPPSU -5vinylon	CSPPSU -10vinylon	CSPPSU -20vinylon
Tensile strength (MPa)	48	27	45	40
Tensile elongation (%) (break)	74	26	48	55
Flexural modulus (MPa) *	757	781	759	548

* Δstress/Δstrain.

3.3. *Proton Conductivities of the CSPPSU-vinylon Membranes*

Polymer electrolyte membranes for high-performance fuel cells require high proton conductivities of >0.01 S/cm, from low to high temperatures and high to low humidification [1]. The conductivity of the CSPPSU-vinylon membranes due to the difference in the amount of PVA added—which is the average value of the error bars of the data obtained after three measurements with varying humidity, at cell temperatures of 40 and 120 °C—is shown in Figure 5. Table 5 shows the physicochemical properties of the CSPPSU-vinylon membranes, depending on the amount of PVA added. The IEC value of the SPPSU polymer was 3.8 meq/g, and the IEC value of the SPPSU-PVA composite polymer was assumed to be equivalent to the IEC value of the SPPSU polymer. Then, the crosslinking degree ($D_{crosslink}$) of the CSPPSU-vinylon membranes was calculated. Moreover, the chemical stability of the membrane was determined using Fenton's reagent (3 wt% H_2O_2 + 2 ppm Fe (II), 80 °C, 1 h). Since vinylon has a high chemical resistance, we thought that the chemical stability of CSPPSU [19] would be improved by incorporating vinylon. However, as shown in Table 5, there was little improvement. With Fenton's reagent, the CSPPSU membrane was radically attacked from the edge, but the CSPPSU-10vinylon membrane was attacked from the inside of the membrane, generating a hole. It is thought that the SPPSU part is selectively vulnerable to attack.

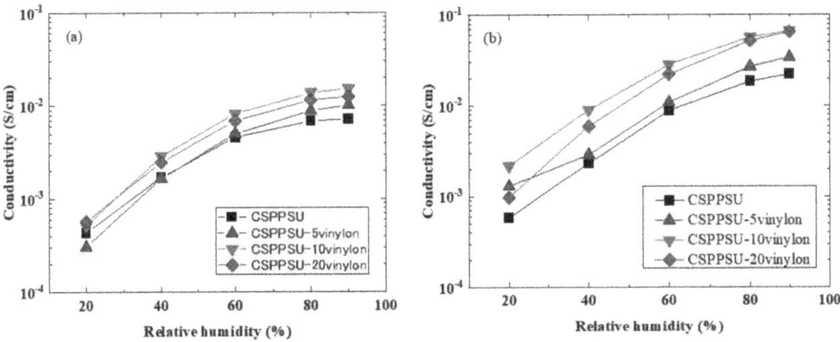

Figure 5. Proton conductivities of the CSPPSU-vinylon membranes vs. the relative humidity at (**a**) 40 and (**b**) 120 °C.

Table 5. Physicochemical and conductivity properties of the CSPPSU-vinylon membranes.

Sample Name			CSPPSU	CSPPSU -5vinylon	CSPPSU -10vinylon	CSPPSU -20vinylon
IEC (meq/g)			2	2.2	2.1	2.1
W.U. (%)			43	38	66	36
λ			11.9	9.6	17.5	9.5
$D_{crosslink}$ (%)			47.3	42.1	44.7	44.7
$R_{oxidation}$ (%)			91–99	_a	81–99	_a
Conductivity (mS/cm)	40 °C	20% RH	0.43	0.31	0.52	0.57
		90% RH	7.23	10.12	15.23	12.43
	80 °C	20% RH	0.7	0.7	0.9	1.1
		90% RH	18	36	56	55
	120 °C	20% RH	0.6	1.3	2.2	1.0
		90% RH	22	34	66	65

_a: disassembly.

On the other hand, the conductivity of the CSPPSU-vinylon electrolyte membrane increased with increases in the temperature and humidity. The conductivity of the CSPPSU-vinylon membrane was higher than that of the CSPPSU membrane. In particular, the CSPPSU-10vinylon membrane had a higher conductivity than the other membranes. The diffusion of protons in the electrolyte membrane depended on the concentration and proton mobility of the sulfonic acid groups (-SO_3H) in the electrolyte membrane, and became faster as the temperature and humidity were increased. In addition, the nanostructure (conduction path) of the electrolyte membrane was greatly affected. The homogeneity of the CSPPSU-10vinylon membrane was better than that of the other membranes (Figure 4). As shown in Table 5, changes in the IEC values of the CSPPSU-vinylon membrane due to the difference in the amount of PVA added, was small and slightly higher than those of the CSPPSU membrane. However, the water content and the number of water molecules per sulfonic acid group (λ) of the CSPPSU-10vinylon membrane, were higher than those of the other crosslinked membranes. These differences contributed to the high proton conductivity of the CSPPSU-10vinylon membrane. The degree of crosslinking of the CSPPSU-vinylon membrane was 42%–45%. It is possible that the hydroxy groups (-OH) of vinylon in the CSPPSU-vinylon membrane contributed to the proton transfer. From the above, it is clear that the conduction mechanism of the SPPSU-vinylon composite membrane is very complicated.

3.4. Fuel Cell Properties using CSPPSU-vinylon Membranes

The performance of fuel cells depends not only on the ionic conductivity of the electrolyte membrane, but also on the interfaces between the electrode layers (catalyst, carbon, ionomer) and between the membrane and the electrode layer [45,49]. Here, the electrode layer and the MEA (membrane electrode assembly) were placed under the same conditions, and only the electrolyte membrane was different. In addition, the measurement conditions for obtaining the current-voltage (I-V) characteristics were the same. The I-V characteristics were evaluated at a cell temperature of 80 °C, and humidities of 100% RH and 60% RH. Figure 6 shows I-$V_{ir\ free}$ and I-iR loss characteristics, evaluated using CSPPSU-10vinylon, CSPPSU, and Nafion212 membranes. The resistance of the single cell using the CSPPSU-10vinylon membrane was higher than that of the Nafion212 membrane, and lower than that of the CSPPSU membrane (Table 6). This tendency in the level of conductivity is the same as that using only the electrolyte membrane (Figure 5). Moreover, the I-$V_{iR\ free}$ characteristics showed the same tendency as the resistance characteristics of the unit cell. On the other hand, when using the CSPPSU-10vinylon membrane, a current of 1.5 A/cm^2 or more was obtained. In the case of

Nafion212 and the CSPPSU membranes, the current was less than 1.5 A/cm^2 at 100% RH, and less than 1 A/cm^2 at 60% RH. Under high humidification conditions, when a high current was applied, flooding occurred on the cathode side, and the voltage tended to drop sharply. Moreover, under low humidification conditions, the membrane resistance increased due to the drying of the membrane, making it difficult to obtain a high current. However, when the CSPPSU-10vinylon membrane was used, a high current without a sharp drop in voltage was obtained, under both high and low humidity conditions. The CSPPSU-vinylon membrane, therefore, seems to have an excellent water treatment ability. These results suggest that the CSPPSU-vinylon membrane would be suitable for thin membrane applications.

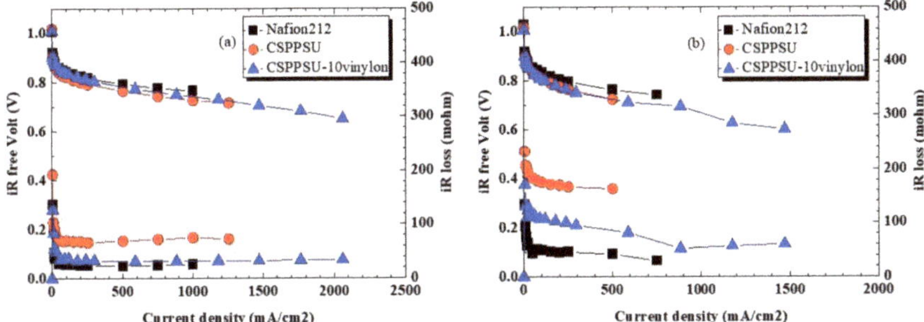

Figure 6. Current-voltage (I-V) properties of CSPPSU-10vinylon, CSPPSU, and Nafion212 membranes: (**a**) I-$V_{iR\ free}$ at 80 °C, 100% RH and (**b**) I-$V_{iR\ free}$ at 80 °C, 60% RH.

Table 6. I-V and H_2 crossover data for single cells using the CSPPSU-10vinylon, CSPPSU, and Nafion212 membranes.

	80 °C, 100% RH			80 °C, 60% RH	
	OCV (V)	**iR Loss (mohm) @ 1 A/cm^2**	**H_2 Crossover (mA/cm^2) @ 0.4 V**	**OCV (V)**	**iR Loss (mohm) @ 1 A/cm^2**
CSPPSU	1.020	73	0.085	1.018	161
CSPPSU-10vinylon	1.010	69	0.245	1.008	119
Nafion212	1.005	24	1.24	1.029	42

Figure 7 shows the hydrogen crossover characteristics of the CSPPSU-10vinylon, CSPPSU, and Nafion212 membranes. The crossover properties of the CSPPSU-10vinylon membrane are five times lower than those of the Nafion212 membrane, and three times higher than those of the CSPPSU membrane (Table 6). We can assume that the crosslinking of SPPSU with vinylon increases the conduction paths (volume fraction) in the nanophase, and hydrogen crossover is higher than that of the CSPPSU membrane.

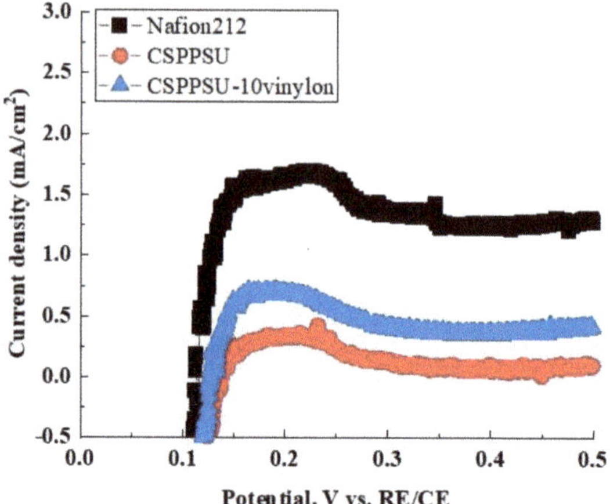

Figure 7. Hydrogen crossover properties of CSPPSU-10vinylon, CSPPSU, and Nafion212 membranes at 80 °C and 100% RH.

4. Conclusions

We focused on improving the performance of CSPPSU membranes with hydrocarbon-based SPPSU polymers, as an alternative electrolyte to fluoropolymer electrolytes. To improve the conductivity and I-V performance properties of CSPPSU membranes, SPPSU and PVA were crosslinked, and CSPPSU-vinylon membranes were synthesized by the formalization of PVA, and compared with Nafion212 and CSPPSU membranes. The conductivities of the CSPPSU-vinylon membranes were higher than those of the CSPPSU membrane. From the results of the fuel cell evaluation, higher current densities than those of Nafion212 and CSPPSU membranes were obtained under both high and low humidification conditions. This is due to the effects of vinylon, and it is thought that the CSPPSU-vinylon membrane has excellent water retention under low humidification conditions. Furthermore, the hydrogen gas crossover properties are lower than those of Nafion212. In other words, the CSPPSU-vinylon membrane would be useful for thin membrane applications.

Author Contributions: Conceptualization, J.-D.K.; Data curation, J.-D.K., S.M. and K.T.; Formal analysis, J.-D.K.; Writing—original draft, J.-D.K.; Writing—review & editing, J.-D.K. All authors have read and agreed to the published version of the manuscript.

Funding: This research received no external funding.

Acknowledgments: This work was partially supported by the MEXT Program for the Development of Environmental Technology, using Nanotechnology from Ministry of Education, Culture, Sports, Science and Technology, Japan.

Conflicts of Interest: The authors declare no conflict of interest.

References

1. Mauritz, K.A.; Moore, R.B. State of Understanding of Nafion. *Chem. Rev.* **2004**, *104*, 4535–4585. [CrossRef]
2. Yandrasits, M.; Lindell, M.; Schaberg, M.; Kurkowski, M. Increasing Fuel Cell Efficiency by Using Ultra-Low Equivalent Weight Ionomers. *Electrochem. Soc. Interface* **2017**, *26*, 49–53. [CrossRef]
3. Matsushita, S.; Kim, J.-D. Organic solvent-free preparation of electrolyte membranes with high proton conductivity using aromatic hydrocarbon polymers and small cross-linker molecules. *Solid State Ionics* **2018**, *316*, 102–109. [CrossRef]
4. Tashvigh, A.A.; Luo, L.; Chung, T.-S.; Weber, M.; Maletzko, C. A novel ionically cross-linked sulfonated polyphenylsulfone (sPPSU) membrane for organic solvent nanofiltration (OSN). *J. Membr. Sci.* **2018**, *545*, 221–228. [CrossRef]

5. Zhang, Y.; Kim, J.-D.; Miyatake, K. Effect of thermal crosslinking on the properties of sulfonated poly(phenylene sulfone)s as proton conductive membranes. *J. Appl. Polym. Sci.* **2016**, *133*, 1–8. [CrossRef]
6. Kim, J.-D.; Ghil, L.-J. Annealing effect of highly sulfonated polyphenylsulfone polymer. *Int. J. Hydrog. Energy* **2016**, *41*, 11794–11800. [CrossRef]
7. Lee, H.; Han, M.; Choi, Y.-W.; Bae, B. Hydrocarbon-based polymer electrolyte cerium composite membranes for improved proton exchange membrane fuel cell durability. *J. Power Sources* **2015**, *295*, 221–227. [CrossRef]
8. Takamuku, S.; Wohlfarth, A.; Manhart, A.; Räder, P.; Jannasch, P. Hypersulfonated polyelectrolytes: Preparation, stability and conductivity. *Polym. Chem.* **2015**, *6*, 1267–1274. [CrossRef]
9. Krishnan, N.N.; Henkensmeier, D.; Jang, J.H.; Hink, S.; Kim, H.-J.; Nam, S.W.; Lim, T.-H. Locally confined membrane modification of sulfonated membranes for fuel cell application. *J. Membr. Sci.* **2014**, *454*, 174–183. [CrossRef]
10. Yu, D.M.; Yoon, S.; Kim, T.H.; Lee, J.Y.; Lee, J.; Hong, Y.T. Properties of sulfonated poly(arylene ether sulfone)/electrospun nonwonen polyacrylonitrile composite membrane for proton exchange membrane fuel cells. *J. Membr. Sci.* **2013**, *446*, 212–219. [CrossRef]
11. Takamuku, S.; Jannasch, P. Properties and degradation of hydrocarbon fuel cell membranes: A comparative study of sulfonated poly(arylene ether sulfone)s with different positions of the acid groups. *Polym. Chem.* **2012**, *3*, 1202–1214. [CrossRef]
12. Wen, P.; Zhong, Z.; Li, L.; Zhang, A.; Li, X.; Lee, M.-H. Photocrosslinking of sulfonated poly(arylene ether sulfone) in a swollen state. *J. Mater. Chem.* **2012**, *22*, 22242–22249. [CrossRef]
13. Chen, S.; Zhang, X.; Chen, K.; Endo, N.; Higa, M.; Okamoto, K.-I.; Wang, L. Cross-linked miscible blend membranes of sulfonated poly(arylene ether sulfone) and sulfonated polyimide for polymer electrolyte fuel cell applications. *J. Power Sources* **2011**, *196*, 9946–9954. [CrossRef]
14. Feng, S.; Shang, Y.; Xie, X.; Wang, Y.; Xu, J. Synthesis and characterization of crosslinked sulfonated poly(arylene ether sulfone) membranes for DMFC applications. *J. Membr. Sci.* **2009**, *335*, 13–20. [CrossRef]
15. Di Vona, M.L.; Luchetti, L.; Spera, G.P.; Sgreccia, E.; Knauth, P. Synthetic strategies for the preparation of proton-conducting hybrid polymers based on PEEK and PPSU for PEM fuel cells. *Comptes Rendus Chim.* **2008**, *11*, 1074–1081. [CrossRef]
16. Xing, D.; Kerres, J.A. Improved performance of sulfonated polyarylene ethers for proton exchange membrane fuel cells. *Polym. Adv. Technol.* **2006**, *17*, 591–597. [CrossRef]
17. Karlsson, L.E.; Jannasch, P. Polysulfone ionomers for proton-conducting fuel cell membranes. *Electrochim. Acta* **2005**, *50*, 1939–1946. [CrossRef]
18. Dyck, A.; Fritsch, D.; Nunes, S.P. Proton-conductive membranes of sulfonated polyphenylsulfone. *J. Appl. Polym. Sci.* **2002**, *86*, 2820–2827. [CrossRef]
19. Kim, J.-D.; Ohira, A.; Nakao, H. Chemically Crosslinked Sulfonated Polyphenylsulfone (CSPPSU) Membranes for PEM Fuel Cells. *Membranes* **2020**, *10*, 31. [CrossRef]
20. Nor, N.A.M.; Nakao, H.; Othman, M.H.D.; Kim, J.-D. Crosslinked carbon nanodots with highly sulfonated polyphenylsulfone as proton exchange membrane for fuel cell applications. *Int. J. Hydrog. Energy* **2020**, *45*, 9979–9988. [CrossRef]
21. Di Vona, M.L.; Alberti, G.; Sgreccia, E.; Casciola, M.; Knauth, P. High performance sulfonated aromatic ionomers by solvothermal macromolecular synthesis. *Int. J. Hydrog. Energy* **2012**, *37*, 8672–8680. [CrossRef]
22. Chen, J.; Zhai, M.; Asano, M.; Huang, L.; Maekawa, Y. Long-term performance of polyetheretherketone-based polymer electrolyte membrane in fuel cells at 95 °C. *J. Mater. Sci.* **2009**, *44*, 3674–3681. [CrossRef]
23. Goto, K.; Rozhanskii, I.; Yamakawa, Y.; Ohtsuki, T.; Naito, Y. Development of aromatic polymer electrolyte membrane with high conductivity and durability for fuel cell. *JSR Tech. Rev.* **2009**, *116*, 1–11. [CrossRef]
24. Di Vona, M.L.; Sgreccia, E.; Licoccia, S.; Alberti, G.; Tortet, L.; Knauth, P. Analysis of temperature-promoted and solvent-assisted cross-linking in sulfonated poly (ether ether ketone)(SPEEK) proton-conducting membranes. *J. Phys. Chem. B* **2009**, *113*, 7505–7512. [CrossRef]
25. Vona, M.L.D.; Sgreccia, E.; Licoccia, S.; Khadhraoui, M.; Denoyel, R.; Knauth, P. Composite proton-conducting hybrid polymers: Water sorption isotherms and mechanical properties of blends of sulfonated PEEK and substituted PPSU. *Chem. Mater.* **2008**, *20*, 4327–4334. [CrossRef]
26. Di Vona, M.L.; Marani, D.; D'Ottavi, C.; Trombetta, M.; Traversa, E.; Beaurroies, I.; Knauth, P.; Lioccia, S. A simple new route to covalent organic/inorganic hybrid proton exchange polymer membranes. *Chem. Mater.* **2006**, *18*, 69–75. [CrossRef]

27. Harrison, W.L.; Hickner, M.; Kim, Y.S.; McGrath, J.E. Poly(Arylene Ether Sulfone) Copolymers and Related Systems from Disulfonated Monomer Building Blocks: Synthesis, Characterization, and Performance—A Topical Review. *Fuel Cells* **2005**, *5*, 201–212. [CrossRef]
28. Xing, D.M.; Yi, B.L.; Liu, F.Q.; Fu, Y.Z.; Zhang, H.M. Characterization of sulfonated poly(etheretherketone)/polytetrafluoroethylene composite membranes for fuel cell applications. *Fuel Cells* **2005**, *5*, 406–411. [CrossRef]
29. Xing, P.; Robertson, G.P.; Guiver, M.D.; Mikhailenko, S.D.; Wang, K.; Kaliaguine, S. Synthesis and characterization of sulfonated poly(etheretherketone) for proton exchange membranes. *J. Membr. Sci.* **2004**, *229*, 95–106. [CrossRef]
30. Mikhailenko, S.D.; Wang, K.; Kaliaguine, S.; Xing, P.; Robertson, G.P.; Guiver, M.D. Proton conducting membranes based on cross-linked sulfonated poly(etheretherketone) (SPEEK). *J. Membr. Sci.* **2004**, *233*, 93–99. [CrossRef]
31. Li, L.; Zhang, J.; Wang, Y. Sulfonated poly(etheretherketon) membranes for direct methanol fuel cell. *J. Membr. Sci.* **2003**, *226*, 159–167. [CrossRef]
32. Feng, S.; Pang, J.; Yu, X.; Wang, G.; Manthiram, A. High-Performance Semicrystalline Poly(ether ketone)-Based Proton Exchange Membrane. *ACS Appl. Mater. Interfaces* **2017**, *9*, 24527–24537. [CrossRef] [PubMed]
33. Ding, X.; Liu, Z.; Hua, M.; Kang, T.; Li, X.; Zhang, Y. Poly(ethylene glycol) crosslinked sulfonated polysulfone composite membranes for forward osmosis. *J. Appl. Polym. Sci.* **2016**, *133*, 1–8. [CrossRef]
34. Yu, J.; Dong, C.; Liu, J.; Li, C.; Fang, J.; Guan, R. Crosslinked sulfonated poly (bis-A)-sulfones as proton exchange membrane for PEM fuel cell application. *J. Mater. Sci.* **2010**, *45*, 1017–1024. [CrossRef]
35. Lafitte, B.; Karlsson, L.E.; Jannasch, P. Sulfophenylation of Polysulfones for Proton-Conducting Fuel Cell Membranes. *Macromol. Rapid Commun.* **2002**, *23*, 896–900. [CrossRef]
36. Lufrano, F.; Gatto, I.; Staiti, P.; Antonucci, V.; Passalacqua, E. Sulfonated polysulfone ionomer membranes for fuel cells. *Solid State Ionics* **2001**, *145*, 47–51. [CrossRef]
37. Kerres, J. Development of ionomer membranes for fuel cells. *J. Membr. Sci.* **2001**, *185*, 3–27. [CrossRef]
38. Lufrano, F.; Squadrito, G.; Patti, A.; Passalacqua, E. Sulfonated polysulfone as promising membranes for polymer electrolyte fuel cells. *J. Appl. Polym. Sci.* **2000**, *77*, 1250–1257. [CrossRef]
39. Schuster, M.; Kreuer, K.-D.; Andersen, H.T.; Maier, J. Sulfonated Poly(phenylene sulfone) Polymers as Hydrolytically and Thermooxidatively Stable Proton Conducting Ionomers. *Macromolecules* **2007**, *40*, 598–607. [CrossRef]
40. Miyake, J.; Taki, R.; Mochizuki, T.; Shimizu, R.; Akiyama, R.; Uchida, M.; Miyatake, K. Design of flexible polyphenylene proton-conducting membrane for next-generation fuel cells. *Sci. Adv.* **2017**, *3*, eaao0476. [CrossRef]
41. Donnadio, A.; Casciola, M.; Di Vona, M.L.; Tamilvanan, M. Conductivity and hydration of sulfonated polyethersulfone in the range 70–120 °C: Effect of temperature and relative humidity cycling. *J. Power Sources* **2012**, *205*, 145–150. [CrossRef]
42. Kim, J.-D.; Donnadio, A.; Jun, M.-S.; Di Vona, M.L. Crosslinked SPES-SPPSU membranes for high temperature PEMFCs. *Int. J. Hydrog. Energy* **2013**, *38*, 1517–1523. [CrossRef]
43. Yin, Y.; Suto, Y.; Sakabe, T.; Chen, S.; Hayashi, S.; Mishima, T.; Yamada, O.; Tanaka, K.; Kita, H.; Okamoto, K.-I. Water Stability of Sulfonated Polyimide Membranes. *Macromolecules* **2006**, *39*, 1189–1198. [CrossRef]
44. Asano, N.; Aoki, M.; Suzuki, S.; Miyatake, K.; Uchida, H.; Watanabe, M. Aliphatic/Aromatic Polyimide Ionomers as a Proton Conductive Membrane for Fuel Cell Applications. *J. Am. Chem. Soc.* **2006**, *128*, 1762–1769. [CrossRef] [PubMed]
45. Aoki, M.; Asano, N.; Miyatake, K.; Uchida, H.; Watanabe, M. Durability of Sulfonated Polyimide Membrane Evaluated by Long-Term Polymer Electrolyte Fuel Cell Operation. *J. Electrochem. Soc.* **2006**, *153*, A1154–A1158. [CrossRef]
46. Wu, D.; Wu, L.; Woo, J.-J.; Yun, S.-H.; Seo, S.-J.; Xu, T.; Moon, S.-H. A simple heat treatment to prepare covalently crosslinked membranes from sulfonated poly(2,6-dimethyl-1,4-phenylene oxide) for application in fuel cells. *J. Membr. Sci.* **2010**, *348*, 167–173. [CrossRef]
47. Robert, F.S.; Litt, M.H. Proton Conducting Polymers Used as Membranes. US Patent Number 5525436, 11 June 1996.

48. Kerres, J.; Ullrich, A.; Meier, F.; Häring, T. Synthesis and characterization of novel acid—base polymer blends for application in membrane fuel cells. *Solid State Ionics* **1999**, *125*, 243–249. [CrossRef]
49. Lee, H.-F.; Killer, M.; Britton, B.; Wu, Y.; Nguyen, H.-D.; Iojoiu, C.; Holdcroft, S. Fuel Cell Catalyst Layers and Membrane-Electrode Assemblies Containing Multiblock Poly(arylene ether sulfones) Bearing Perfluorosulfonic Acid Side Chains. *J. Electrochem. Soc.* **2018**, *165*, F891–F897. [CrossRef]

Article

Electrospun Anion-Conducting Ionomer Fibers—Effect of Humidity on Final Properties

Manar Halabi [1], Meirav Mann-Lahav [1], Vadim Beilin [1], Gennady E. Shter [1], Oren Elishav [1,2], Gideon S. Grader [1,2,*] and Dario R. Dekel [1,2,*]

1 The Wolfson Department of Chemical Engineering, Technion–Israel Institute of Technology, Haifa 3200003, Israel; manarh@campus.technion.ac.il (M.H.); meiravml@technion.ac.il (M.M.-L.); vadimbe@technion.ac.il (V.B.); shter@technion.ac.il (G.E.S.); orene@campus.technion.ac.il (O.E.)

2 The Nancy & Stephan Grand Technion Energy Program (GTEP), Technion, Israel Institute of Technology, Haifa 3200003, Israel

* Correspondence: grader@technion.ac.il (G.S.G.); dario@technion.ac.il (D.R.D.)

Received: 3 April 2020; Accepted: 27 April 2020; Published: 1 May 2020

Abstract: Anion-conducting ionomer-based nanofibers mats are prepared by electrospinning (ES) technique. Depending on the relative humidity (RH) during the ES process (RH_{ES}), ionomer nanofibers with different morphologies are obtained. The effect of relative humidity on the ionomer nanofibers morphology, ionic conductivity, and water uptake (WU) is studied. A branching effect in the ES fibers found to occur mostly at RH_{ES} < 30% is discussed. The anion conductivity and WU of the ionomer electrospun mats prepared at the lowest RH_{ES} are found to be higher than in those prepared at higher RH_{ES}. This effect can be ascribed to the large diameter of the ionomer fibers, which have a higher WU. Understanding the effect of RH during the ES process on ionomer-based fibers' properties is critical for the preparation of electrospun fiber mats for specific applications, such as electrochemical devices.

Keywords: electrospinning; fibrous morphology; relative humidity; polymer fibers; ionomer

---

1. Introduction

Electrospinning (ES) is broadly applied to generate nanofibers from a wide range of materials, including polymers, metals, ceramics, and composites [1–5]. This technique [6–10] allows control of the fibers' morphology and diameter, which play important roles in their final applications. Solution properties such as precursor concentration, polymer molecular weight, viscosity, solvent characteristics as well as process conditions greatly affect the electrospun fibers [7,8,11–14]. Such nanofibers have potential applications in the biomedical [15,16], energy [17,18], and other fields. The electrospinning process often involves usage of organic solvents, which are toxic, expensive, and considered environmentally unfriendly. This motivated the researchers to develop the "green process", which is a more eco-friendly ES process that uses aqueous polymer solutions [19,20]. In addition, the high viscosities of the polymer solutions require using low polymer concentrations, thereby limiting the utilization of the ES process due to the large amounts of organic solvents needed. The "green process" overcomes these limitations, allowing higher polymer concentrations in water for more effective and productive ES procedure [21].

The influence of environmental conditions during electrospinning on the morphology of polymer fibers has been investigated [11,22,23]. For example, the relative humidity during ES, RH_{ES}, plays a critical role for the formation of "porous morphology" [24]. Casper et al. [12] studied the effect of increasing RH_{ES} and varying polystyrene (PS) molecular weight (MW) on the pore size distribution of the electrospun fibers. They found that raising the RH_{ES} increases the number of pores and their diameter. Higher MW leads to larger pores on the fibers' surface. The effect of changing RH_{ES} and solvents ratios on the surface morphology of electrospun PS fibers was reported elsewhere [25,26]. A study demonstrated the effects of solution's viscosity and RH_{ES} on fibers' morphology and claimed that monitoring these two parameters facilitated fibers' architecture control, which changed from beaded fibers to smooth uniform fibers [23]. In addition to examining the external morphology, Pai et al. [27] presented porous structures within electrospun PS fibers while electrospinning at RH_{ES} in the 24–43% range. Based on TEM analysis, the void fraction of electrospun PS in DMF was about 30%. It was shown that the internal pores have a significant effect on the mechanical, optical, as well as electrical properties of the fibers.

Although electrospinning of (inert) polymers has been extensively reported, this process was scarcely studied for the case of ionic polymers, usually called ionomers [28–37]. While some proton-conducting ionomers have been the base of first attempts to produce electrospun proton-conducting fiber mats [38–43], only a few studies can be found on electrospun anion-conducting fiber mats [17,44,45].

In this work, we investigate the electrospinning process of an anion-conducting ionomeric material. We specifically focus on an interesting branching phenomenon that is observed in anion-conducting fibers that were electrospun under 30% RH_{ES}. According to a study by Yarin et al. [46], branching occurs when the static undulations of a cylindrical jet become unstable at the sites of the highest local curvature, where secondary jet branches are ejected from the primary jet. This phenomenon was also explained by Tan [47]. The branching effect had been demonstrated on polymer fibers [48] and piezoelectric fibers [49]; however, all the studies mentioned above were done using nonconducting polymers. The branching effect can be leveraged to drastically increase the surface area of fiber products; nevertheless, it is unclear what governs this effect in conducting polymers and whether they are as sensitive to the environmental conditions during the electrospinning process as nonconductive polymers. Moreover, due to the ionic character of the anion-conducting ionomers, it is especially interesting to investigate the effect of the ES process on their final properties.

The goal of this article is to investigate the effect of RH_{ES} on the electrospun anion-conducting polymer nanofibers' morphology as well as fibers' mat properties. The effect of RH_{ES} on the branching of anion-conducting polymer is presented and a qualitative model explaining this effect is suggested. Understanding the relationship between RH_{ES} and the structure, WU, and anion conductivity of the electrospun ionomer nanofibers would enable us to achieve fiber mats with tailored and improved properties for their numerous applications in electrochemical devices. Such fiber mats can be used, for instance, in advanced anion exchange membrane fuel cells (AEMFCs) [50–54] as a catalyst bed as well as a membrane.

2. Experimental

Anion-conducting ionomeric material (FAA-3 in its Br^- form) was purchased from Fumatech BWT GmbH, Germany. FAA-3 is a quaternary ammonium functionalized aromatic PPO (Poly (p-phenylene oxide))-based polymer. The solvent used for preparing the ES precursor solutions was N,N-dimethylformamide (DMF) (Bio-Lab ltd., Jerusalem, Israel). Precursor solutions were prepared by dissolving the anion-conducting ionomer in DMF with ionomer concentration of 30 wt %. The solutions were magnetically stirred for 2 h at room temperature until complete dissolution of the ionomer.

The ionomer solutions were electrospun in an electrospinning machine—NS24, (Inovenso, Istanbul, Turkey), in an environmentally controlled chamber. Precursors were loaded into a 5-mL syringe and fed to the ES system at a flow rate of 0.3 mL h^{-1} for 6 h. The tip to collector distance was 14.5 cm. A positive charge of 20 kV was applied on the needle, and a negative charge of 3 kV was applied to a flat collector. The flat collector size is 130 mm by 370 mm, it moved back and forth 60 mm along the long axis at a speed of 10 mm s^{-1} to ensure fiber deposition homogeneity. The chamber's temperature was set constant at 24–25 °C, and the RH_{ES} was varied in the range of 20%–50%.

The obtained ionomeric mats were dried in a vacuum oven at 40 °C for 16 h to remove the residual solvent. The morphology of the nanofibers was characterized by high-resolution scanning electron microscopy (HR-SEM, Ultra Plus, Zeiss, Switzerland) at a magnification range of ×1000–10,000.

The diameter of the electrospun fibers was obtained by HR-SEM images analysis of three different areas on ionomer mats deposited on carbon conductive tape. The fibers' diameter distribution was calculated by a statistical analysis of at least 30 data points from each sample.

The WU of the electrospun ionomer samples was measured using a VTI SA + instrument (TA Instruments, New Castle, DE, USA), using protocols detailed elsewhere [55–58]. In brief, each electrospun ionomer sample was dried in situ for a maximum of 60 min at 50 °C and RH close to 0%. After that, the temperature was set constant at 40 °C and the RH was then raised from 10% to 90% in intervals of 20%. Each RH step was maintained until the sample weight reached equilibrium (WU changes smaller than 0.001 wt % in 5 min). The temperature during the WU measurements was kept at 40 °C. Values of WU were calculated using Equation (1) [55,59], where the "wet" and "dry" weight ($W_{(wet)}$ and $W_{(dry)}$, respectively) were measured at the end of each equilibrium step and at the end of the initial drying step, respectively:

$$WU = \frac{W(wet) - W(dry)}{W(dry)} \times 100\%. \quad (1)$$

The WU kinetics were also determined, by measuring the mass change of the ionomer mat electrospun sample as a function of time at every RH step. The characteristic time constant, τ, was calculated by fitting the experimental data with the following equation [55,60]:

$$\frac{W_t - W_0}{W_\infty - W_0} = \frac{M_t}{M_\infty} \cong 1 - exp\left(-\frac{t}{\tau}\right), \quad (2)$$

where W_t is the mass of sample at time t, W_0 is the mass at the beginning of the RH step, W_∞ is the mass of membrane at equilibrium state, M_t is the mass gain at time t, and M_∞ is the mass gain of the ionomer mat at equilibrium.

Bromide anion conductivity measurements were conducted on the ionomer mats electrospun at different RHs. These measurements were made in a conductivity chamber of an MTS-740 ionic conductometer (Scribner Inc., Southern Pines, NC, USA) using the protocol detailed elsewhere [58]. The bromide anion conductivity was calculated using the ionomer mat resistance measured with a 4-point probe cell in a sealed, thermally insulated chamber under continuous N_2 gas conditioned to the desired humidity. The ionomer mat samples were first equilibrated at 40 °C at 90% RH for 1 h, the RH was then decreased from 90% to 10% in intervals of 20%, then back from 10% to 90%, following the procedure reported elsewhere [61]. Resistance values were measured perpendicular to the fiber mat, in the through-plane (TP) direction. The TP resistance, R, was measured by impedance spectroscopy using PSM1735 Frequency Response Analyzer (Newtons4th Ltd., Leicester, UK). The TP conductivity was then calculated as [61,62]

$$\sigma_{TP} = \frac{d}{A \cdot R}, \quad (3)$$

where d is the mat thickness, A is the cross-sectional area through which the current passes (0.5 cm^2), and R is the measured resistance.

The velocity of the jet was calculated using the model reported by Ding et al. [63]. This model is based on the mass conversation of the polymer in the fiber. While the jet is elongated under different ES parameters and times, the volume of polymer along the fiber length must equal the volume of polymer that was consumed. Therefore, the model considers the diameter of the jet. The velocity is calculated by

$$v_1 = \frac{Q_s}{S_1 \cdot t}, \quad (4)$$

where Q_s stands for the consumption of spinning solution (mL), S_1 is the cross-sectional area of the fiber (μm^2), and t is the electrospinning time (h).

3. Results and Discussion

3.1. Ionomer Fibers Morphology

The ionomer fibers electrospun at RH_{ES} = 20% were mostly flat belts. This can be rationalized by the formation of an early skin on the fibers due to fast evaporation at RH_{ES} = 20%. The skin prevents uniform fiber shrinkage and eventually collapses into a belt (Figure 1a). A similar effect has been observed and modeled in our work on other fiber materials [64,65]. At RH_{ES} = 30%, a more significant branching effect was observed (Figure 1b); while at RH_{ES} = 40% and 50%, the fibers were smooth and cylindrical (Figure 1c,d) without any visible branching. This observation of the electrospun ionomer branching effect is further confirmed by measuring the fiber diameter distribution (Figure 2).

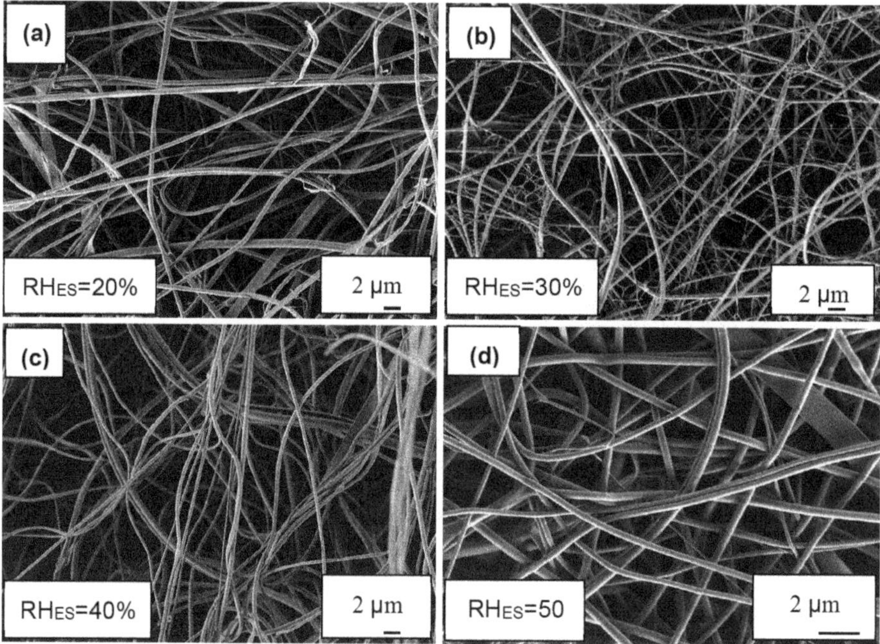

Figure 1. SEM images of ionomer fibers electrospun at RH_{ES} range of 20%–50%. All SEM images taken at ×5000 magnification. (**a**) 20% RH_{ES}, (**b**) 30% RH_{ES}, (**c**) 40% RH_{ES} and (**d**) 50% RH_{ES}.

The diameter distributions at RH_{ES} = 20%, 40%, and 50% are monomodal (Figure 2a,c,d), whereas at RH_{ES} = 30%, a bimodal diameter distribution can clearly be seen in Figure 2b. The diameter of the branches (small fibers) is in the 80–180 nm range.

As RH_{ES} increased from 20% to 30%–50%, the main average fiber diameter decreased from ~600 nm to a constant value of ~400 nm (see Figure 3). The larger width of the fibers electrospun at RH_{ES} = 20% may be due to their flat morphology, where the original cylindrical fiber collapsed into a belt, thus spreading over a larger width. The bimodal diameter distribution at 30% RH_{ES} can be explained by the branching mechanism, due to excess charge in unstable areas on the primary cylindrical jet [48,49]. Although some small ionomer fibers are also observed during ES at RH_{ES} = 20%, the diameter distribution of the electrospun ionomer fibers is not bimodal.

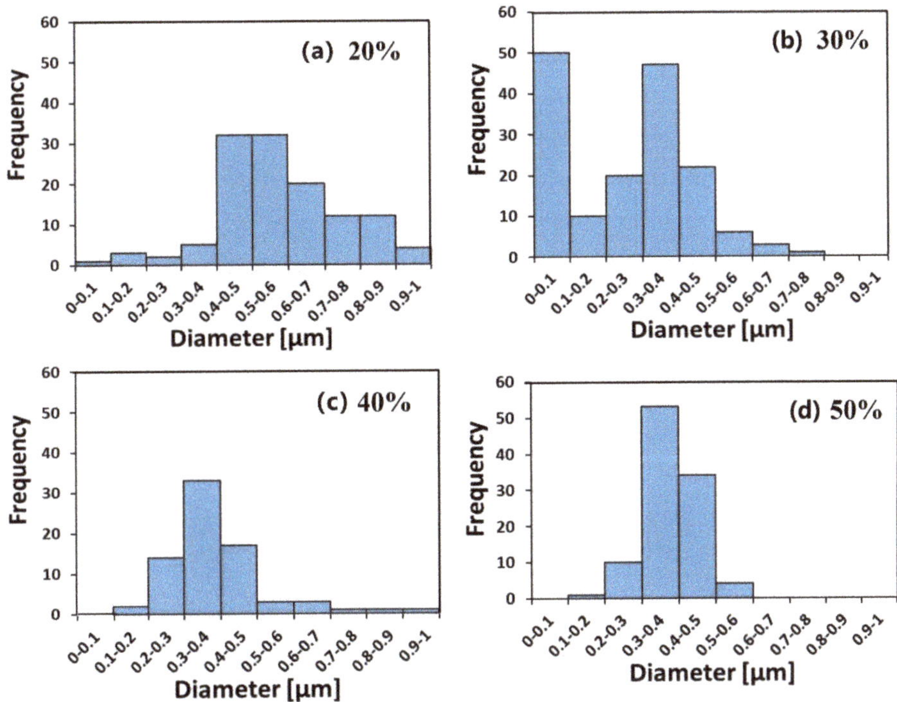

Figure 2. Fibers' diameter distribution of ionomer fibers electrospun under a relative humidity (RH_{ES}) range of 20%–50%. (**a**) 20% RH_{ES}, (**b**) 30% RH_{ES}, (**c**) 40% RH_{ES} and (**d**) 50% RH_{ES}.

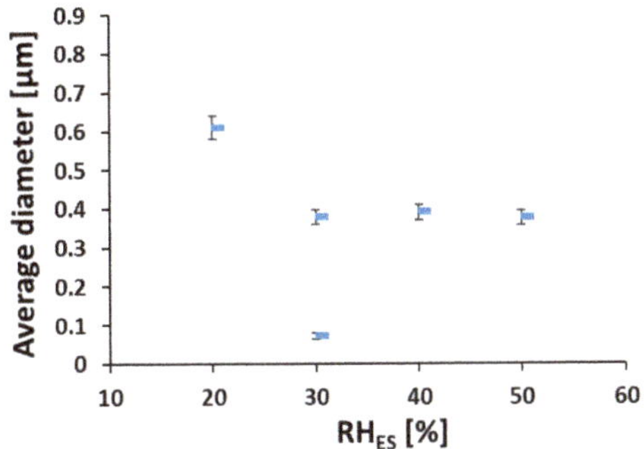

Figure 3. Fibers' average diameter vs. RH_{ES} (calculated from data in Figure 2).

Depending on the hydrophobic or hydrophilic nature of the polymer [66], the RH can either increase or decrease the nanofibers' diameter. When hydrophilic polymers such as polyvinylpyrrolidone and polyethylene oxide are electrospun at higher RH, the polymers solutions solidify more slowly due to slower evaporation rate [67]. This gives rise to a smaller diameter due to further stretching of the fiber. Another effect that plays a role here is that at a higher RH, the electric charges on the fibers' surface can discharge to the surrounding water vapor more easily. Thus, the charge on the fiber is reduced, hence the attraction to the collector and stretching force are reduced, which in turn leads to larger fiber diameters [26]. In the present case, ES at low RH_{ES} (20%) caused fast evaporation of DMF, giving rise mostly to flat fibers with an average width of 600 nm. Above RH_{ES} = 20%, fast evaporation of the solvent is prevented, more stretching occurs, and the cylindrical cross-section is maintained. At 40% and 50% RH_{ES}, the discharging effect becomes more dominant, thus we do not see an additional decrease in fibers' diameter below 400 nm as shown in Figure 4.

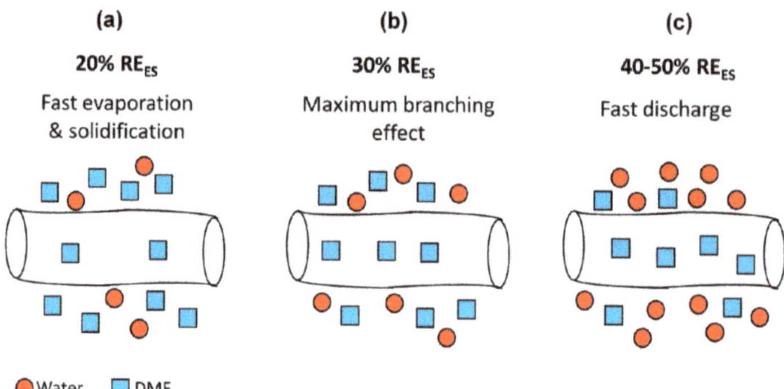

Figure 4. Scheme of electrospun fibers during their flight towards the collector at different RH_{ES}: (**a**) 20%, (**b**) 30%, and (**c**) 40–50%.

The two trends described above, involving solvent evaporation and discharging, occur simultaneously and have opposing effects. The maximal branching phenomena at 30% RH_{ES} has not been observed before. It is not clear why branching is subdued at lower RH and eliminated at higher RH. Baumgarten et al. [68] investigated the effect of surrounding gas during ES on fibers' morphology. Electrospinning in Freon atmosphere gave rise to a branching effect. However, ES in air at identical conditions (15 kV and 75 TCD) did not show any branching. This difference was attributed to the higher breakdown voltage of Freon compared to air. We believe that the lower breakdown voltage of the humid atmosphere during ES at high RH_{ES} (40–50%) causes a discharge of the electric charges on the fibers' surface to the surrounding water vapor. This effect reduces the surface charge concentration at the unstable points; thus, branching does not occur.

3.2. Water Uptake

It can be seen (Figure 5a) that the WU increases with RH for all the ionomer mat samples. Over the whole range of RH values, the ionomer mat sample prepared at RH_{ES} = 20% has higher WU compared with those electrospun at higher RH_{ES}. The ionomer mat samples prepared at RH_{ES} = 30% and 40% have almost the same WU, even though the sample prepared at RH_{ES} = 30% has a bimodal fiber distribution. The reason for this is that the weight fraction of the small diameter segment in the bimodal sample is 1/16 of the larger diameter segment. (The average diameter of the fibers in the bimodal sample are 50 and 400 nm, while the ratio of small vs. large ionomer fibers per unit area is 4). Hence, water absorption on the small fibers does not contribute significantly to the overall WU.

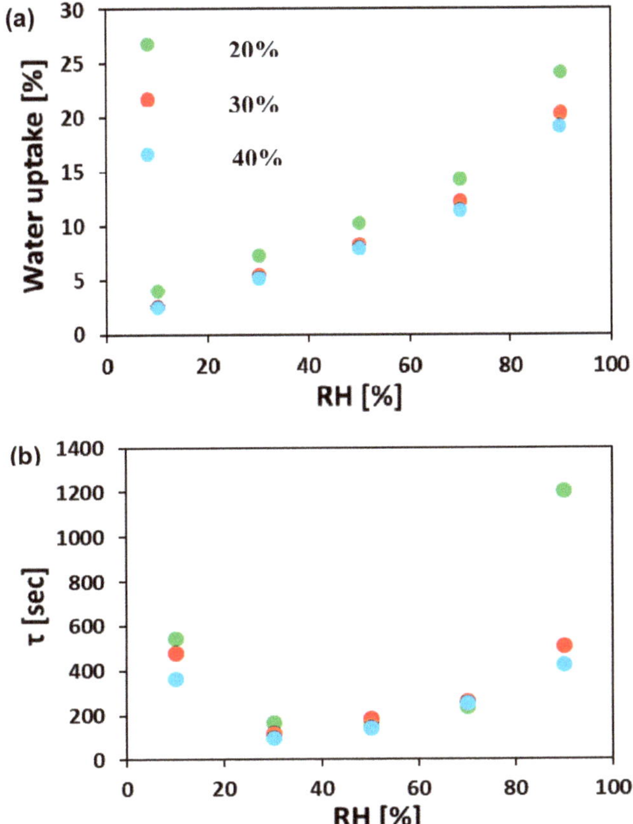

Figure 5. (**a**) Water vapor absorption isotherms and (**b**) characteristic time constant, τ, for electrospun ionomer mats. All tests were done at 40 °C. RH—relative humidity.

The characteristic time constant, τ, was calculated from Equation (2), as described in the experimental section. Figure 5b summarizes the resulting τ for all samples. The electrospun ionomer mats had a relatively large τ at 10% RH (350–550 s). As RH is increased, τ decreases to a minimum at RH (30–50%) and then increases (τ = 400–1200 s) at higher RH. The minimal value of τ in the 30%–50% range of RH indicates that the fastest WU kinetics in these materials occurs in the mid-RH region. This is consistent with the trends found for WU studies of different nonporous anion-conducting membranes [55,64].

The high value of τ at low RH can be rationalized by the presence of smaller water content in the gas phase, thus resulting in a lower flux of water molecules towards the surface. The faster WU kinetics (lower values of τ) in the mid-range RH (30–50% RH) can be explained by the larger initial water content within the ionomer fibers, which lowers their density and the larger flux of water from the gas phase at high RH. This combination makes it easier for additional water molecules on the surface to diffuse into the ionomer fiber. When the RH is the highest (90%), the fibers are nearly saturated and hence it is more difficult for additional water molecules to diffuse through the fiber, resulting therefore in higher τ values.

3.3. Anion Conductivity

Figure 6 shows the bromide TP anion conductivity at 40 °C as a function of RH for ionomer fibers prepared at different RH_{ES}. As can be seen, Br^- anion conductivity significantly increases with increasing RH. As can be seen, the anion conductivity of the electrospun ionomer mats made at lower RH_{ES} are higher than mats electrospun at larger RH_{ES}. For instance, at 70% RH, the Br^- TP anion conductivity of electrospun mat under 20% RH_{ES} is 6 times higher than that of a mat prepared at 40% RH_{ES}.

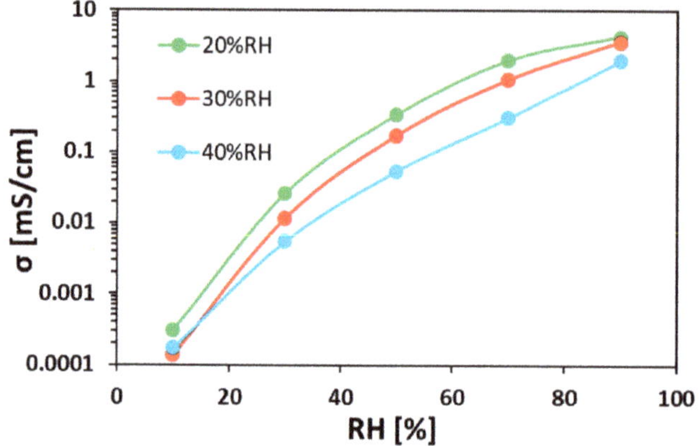

Figure 6. Bromide TP anion conductivity at 40 °C vs. RH for electrospun ionomer fibers prepared at different RH_{ES}.

A clear relation can be concluded from the results in Figures 5a and 6. In both figures, the order is kept: the electrospun ionomer mat at 20% RH_{ES} has the highest WU and highest anion conductivity at RHs above 50%, compared to both mats prepared at 30% and 40% RH_{ES}. This indicates that higher WU and density promote faster ion transfer within the ionomer fibers and between them, which also explains the higher anion conductivity of the mat electrospun at 20% RH_{ES}.

4. Conclusions

In this study, anion-conducting fiber mats were prepared by electrospinning (ES) at different relative humidity (RH_{ES}) values. The ionomer nanofibers' morphology, anion conductivity, and water uptake (WU) were measured. Formation of branched thin fibers was observed in mats prepared at RH_{ES} 20% and 30%. The mechanism of branching as a function of RH_{ES} in ES of anion-conducting material was discussed. It was shown that ionomer fiber diameter distribution is significantly affected by the RH_{ES} up to RH = 40%. Above this value, the effect of RH_{ES} on the fiber diameter distribution was marginal.

The results of WU and anion conductivity measurements showed that the ionomer mat electrospun under the lowest RH_{ES} (20%) gained the highest WU (24%) and TP conductivity (4.4 mS cm^{-1}) at RH = 90%. These results are associated with the large fibers' diameter and their high water capacity, which gives rise to higher anion conductivity.

The preparation of anion-conducting nanofibers prepared by ES process can be attractive for numerous applications including AEMFCs. Therefore, understanding the interplay between process conditions such as relative humidity during electrospinning and the nanofibers' structure, WU, as well as anion conductivity is important for the future design of these devices with tailored properties. This understanding is critical for effective water management within these future devices.

Author Contributions: Data curation, O.E.; investigation, M.H. and M.M.-L.; methodology, V.B. and G.E.S.; supervision, M.M.-L., G.E.S., G.S.G., and D.R.D.; writing—original draft, M.H. and M.M.-L.; writing—review and editing, G.S.G. and D.R.D. All authors have read and agreed to the published version of the manuscript.

Acknowledgments: This work was partially funded by the Nancy & Stephan Grand Technion Energy Program (GTEP); by the Ministry of Science, Technology & Space of Israel through grant No. 3-12948; by the Israel Science Foundation (ISF) [grant No. 1481/17]; by the Russell Berrie Nanotechnology Institute, Technion; by the Ministry of National Infrastructure, Energy and Water Resources of Israel [grant No. 3-13671]. The authors would also like to acknowledge the financial support of Melvyn & Carolyn Miller Fund for Innovation, as well as the support of Planning & Budgeting Committee/ISRAEL Council for Higher Education (CHE) and Fuel Choice Initiative (Prime Minister Office of ISRAEL), within the framework of "Israel National Research Center for Electrochemical Propulsion (INREP)". GSG also acknowledges support from the Gruenbaum chair in Materials Engineering.

Conflicts of Interest: The authors declare no conflict of interest.

References

1. Chronakis, I.S. Novel nanocomposites and nanoceramics based on polymer nanofibers using electrospinning process—A review. *J. Mater. Process. Technol.* **2005**, *167*, 283–293. [CrossRef]
2. Teo, W.E.; Ramakrishna, S. A review on electrospinning design and nanofibre assemblies. *Nanotechnology* **2006**, *17*. [CrossRef] [PubMed]
3. Halperin, V.; Shter, G.E.; Beilin, V.; Grader, G.S. Mesoporous K/Fe–Al–O nanofibers by electrospinning of solution precursors. *J. Mater. Res.* **2015**, *30*, 3142–3150. [CrossRef]
4. Shmueli, Y.; Shter, G.E.; Assad, O.; Haick, H.; Sonntag, P.; Ricoux, P.; Grader, G.S. Structural and electrical properties of single Ga/ZnO nanofibers synthesized by electrospinning. *J. Mater. Res.* **2012**, *27*, 1672–1679. [CrossRef]
5. Elishav, O.; Beilin, V.; Shter, G.E.; Dinner, O.; Halperin, V.; Grader, G.S. Formation of Core-Shell Mesoporous Ceramic Fibers. *J. Am. Ceram. Soc.* **2017**, *100*, 3370–3374. [CrossRef]
6. Cong, Y.; Liu, S.; Chen, H. Fabrication of conductive polypyrrole nanofibers by electrospinning. *J. Nanomater.* **2013**, *2013*, 1–7. [CrossRef]
7. Wang, Z.M. *One-Dimensional Nanostructures*; Springer Science & Business Media: Berlin, Germany, 2008; Volume 3, ISBN 0387741321.
8. Baji, A.; Mai, Y.W.; Wong, S.C.; Abtahi, M.; Chen, P. Electrospinning of polymer nanofibers: Effects on oriented morphology, structures and tensile properties. *Compos. Sci. Technol.* **2010**, *70*, 703–718. [CrossRef]
9. Junoh, H.; Jaafar, J.; Norddin, M.N.A.M.; Ismail, A.F.; Othman, M.H.D.; Rahman, M.A.; Yusof, N.; Salleh, W.N.W.; Ilbeygi, H. A review on the fabrication of electrospun polymer electrolyte membrane for direct methanol fuel cell. *J. Nanomater.* **2015**, *2015*, 4. [CrossRef]
10. Cavaliere, S.; Subianto, S.; Savych, I.; Jones, D.J.; Rozière, J. Electrospinning: Designed architectures for energy conversion and storage devices. *Energy Environ. Sci.* **2011**, *4*, 4761. [CrossRef]
11. Megelski, S.; Stephens, J.S.; Chase, D.B.; Rabolt, J.F. Micro-and nanostructured surface morphology on electrospun polymer fibers. *Macromolecules* **2002**, *35*, 8456–8466. [CrossRef]
12. Casper, C.L.; Stephens, J.S.; Tassi, N.G.; Chase, D.B.; Rabolt, J.F. Controlling surface morphology of electrospun polystyrene fibers: Effect of humidity and molecular weight in the electrospinning process. *Macromolecules* **2004**, *37*, 573–578. [CrossRef]
13. Pisignano, D. *Polymer Nanofibers: Building Blocks for Nanotechnology*; Royal Society of Chemistry: London, UK, 2013; Volume 8, ISBN 1849735743.
14. Ramakrishna, S. *An Introduction to Electrospinning and Nanofibers*; WORLD SCIENTIFIC: Singapore, 2005; ISBN 978-981-256-415-3.
15. Haider, A.; Haider, S.; Kang, I.-K. A comprehensive review summarizing the effect of electrospinning parameters and potential applications of nanofibers in biomedical and biotechnology. *Arab. J. Chem.* **2018**, *11*, 1165–1188. [CrossRef]
16. Pham, Q.P.; Sharma, U.; Mikos, A.G. Electrospinning of polymeric nanofibers for tissue engineering applications: A review. *Tissue Eng.* **2006**, *12*, 1197–1211. [CrossRef] [PubMed]
17. Park, A.M.; Pintauro, P.N. Alkaline Fuel Cell Membranes from Electrospun Fiber Mats. *Electrochem. Solid-State Lett.* **2012**, *15*, B27. [CrossRef]

18. Bajon, R.; Balaji, S.; Guo, S.M. Electrospun Nafion Nanofiber for Proton Exchange Membrane Fuel Cell Application. *J. Fuel Cell Sci. Technol.* **2009**, *6*, 031004. [CrossRef]
19. Sundarrajan, S.; Ramakrishna, S. Green Processing of a Cationic Polyelectrolyte Nanofibers in the Presence of Poly (vinyl alcohol). *Int. J. Green Nanotechnol.* **2011**, *3*, 244–249. [CrossRef]
20. Sridhar, R.; Sundarrajan, S.; Vanangamudi, A.; Singh, G.; Matsuura, T.; Ramakrishna, S. Green processing mediated novel polyelectrolyte nanofibers and their antimicrobial evaluation. *Macromol. Mater. Eng.* **2014**, *299*, 283–289. [CrossRef]
21. Sun, J.; Bubel, K.; Chen, F.; Kissel, T.; Agarwal, S.; Greiner, A. Nanofibers by green electrospinning of aqueous suspensions of biodegradable block copolyesters for applications in medicine, pharmacy and agriculture. *Macromol. Rapid Commun.* **2010**, *31*, 2077–2083. [CrossRef]
22. Huang, L.; Bui, N.; Manickam, S.S.; McCutcheon, J.R. Controlling electrospun nanofiber morphology and mechanical properties using humidity. *J. Polym. Sci. Part B Polym. Phys.* **2011**, *49*, 1734–1744. [CrossRef]
23. Nezarati, R.M.; Eifert, M.B.; Cosgriff-Hernandez, E. Effects of humidity and solution viscosity on electrospun fiber morphology. *Tissue Eng. Part C Methods* **2013**, *19*, 810–819. [CrossRef]
24. Zheng, J.; Zhang, H.; Zhao, Z.; Han, C.C. Construction of hierarchical structures by electrospinning or electrospraying. *Polymer* **2012**, *53*, 546–554. [CrossRef]
25. Park, J.-Y.; Lee, I.-H. Relative humidity effect on the preparation of porous electrospun polystyrene fibers. *J. Nanosci. Nanotechnol.* **2010**, *10*, 3473–3477. [CrossRef] [PubMed]
26. Kim, G.-T.; Lee, J.-S.; Shin, J.-H.; Ahn, Y.-C.; Hwang, Y.-J.; Shin, H.-S.; Lee, J.-K.; Sung, C.-M. Investigation of pore formation for polystyrene electrospun fiber: Effect of relative humidity. *Korean J. Chem. Eng.* **2005**, *22*, 783–788. [CrossRef]
27. Pai, C.-L.; Boyce, M.C.; Rutledge, G.C. Morphology of porous and wrinkled fibers of polystyrene electrospun from dimethylformamide. *Macromolecules* **2009**, *42*, 2102–2114. [CrossRef]
28. Ono, H.; Miyake, J.; Shimada, S.; Uchida, M.; Miyatake, K. Anion exchange membranes composed of perfluoroalkylene chains and ammonium-functionalized oligophenylenes. *J. Mater. Chem. A* **2015**, *3*, 21779–21788. [CrossRef]
29. Melo, L.D.; Mamizuka, E.M.; Carmona-Ribeiro, A.M. Antimicrobial particles from cationic lipid and polyelectrolytes. *Langmuir* **2010**, *26*, 12300–12306. [CrossRef]
30. Huang, T.; He, G.; Xue, J.; Otoo, O.; He, X.; Jiang, H.; Zhang, J.; Yin, Y.; Jiang, Z.; Douglin, J.C. Self-crosslinked blend alkaline anion exchange membranes with bi-continuous phase separated morphology to enhance ion conductivity. *J. Memb. Sci.* **2020**, *597*, 117769. [CrossRef]
31. Siroma, Z.; Ioroi, T.; Fujiwara, N.; Yasuda, K. Proton conductivity along interface in thin cast film of Nafion®. *Electrochem. Commun.* **2002**, *4*, 143–145. [CrossRef]
32. Willdorf-Cohen, S.; Mondal, A.N.; Dekel, D.R.; Diesendruck, C.E. Chemical stability of poly (phenylene oxide)-based ionomers in an anion exchange-membrane fuel cell environment. *J. Mater. Chem. A* **2018**, *6*, 22234–22239. [CrossRef]
33. Karan, K. Interesting Facets of Surface, Interfacial, and Bulk Characteristics of Perfluorinated Ionomer Films. *Langmuir* **2019**, *35*, 13489–13520. [CrossRef]
34. Amel, A.; Gavish, N.; Zhu, L.; Dekel, D.R.; Hickner, M.A.; Ein-Eli, Y. Bicarbonate and chloride anion transport in anion exchange membranes. *J. Memb. Sci.* **2016**, *514*, 125–134. [CrossRef]
35. Maurya, S.; Shin, S.-H.; Kim, Y.; Moon, S.-H. A review on recent developments of anion exchange membranes for fuel cells and redox flow batteries. *RSC Adv.* **2015**, *5*, 37206–37230. [CrossRef]
36. Merle, G.; Wessling, M.; Nijmeijer, K. Anion exchange membranes for alkaline fuel cells: A review. *J. Memb. Sci.* **2011**, *377*, 1–35. [CrossRef]
37. Amel, A.; Smedley, S.B.; Dekel, D.R.; Hickner, M.A.; Ein-Eli, Y. Characterization and chemical stability of anion exchange membranes cross-linked with polar electron-donating linkers. *J. Electrochem. Soc.* **2015**, *162*, F1047–F1055. [CrossRef]
38. Park, J.; Wycisk, R.; Pintauro, P.; Yarlagadda, V.; Van Nguyen, T. Electrospun Nafion®/Polyphenylsulfone composite membranes for regenerative Hydrogen bromine fuel cells. *Materials.* **2016**, *9*, 143. [CrossRef]
39. Choi, J.; Lee, K.M.; Wycisk, R.; Pintauro, P.N.; Mather, P.T. Nanofiber composite membranes with low equivalent weight perfluorosulfonic acid polymers. *J. Mater. Chem.* **2010**, *20*, 6282. [CrossRef]
40. Ballengee, J.B.; Pintauro, P.N. Preparation of nanofiber composite proton-exchange membranes from dual fiber electrospun mats. *J. Memb. Sci.* **2013**, *442*, 187–195. [CrossRef]

41. Sood, R.; Cavaliere, S.; Jones, D.J.; Rozière, J. Electrospun nanofibre composite polymer electrolyte fuel cell and electrolysis membranes. *Nano Energy* **2016**, *26*, 729–745. [CrossRef]
42. Subianto, S.; Giancola, S.; Ercolano, G.; Nabil, Y.; Jones, D.; Rozière, J.; Cavaliere, S. Electrospun nanofibers for low-temperature proton exchange membrane fuel cells. In *Electrospinning for Advanced Energy and Environmental Applications*; Sara, C., Ed.; CRC Press: Boca Raton, FL, USA, 2015; pp. 29–59. ISBN 1482217678.
43. Subianto, S.; Cavaliere, S.; Jones, D.J.; Rozière, J. Effect of side-chain length on the electrospinning of perfluorosulfonic acid ionomers. *J. Polym. Sci. Part A Polym. Chem.* **2013**, *51*, 118–128. [CrossRef]
44. Park, A.M.; Turley, F.E.; Wycisk, R.J.; Pintauro, P.N. Electrospun and Cross-Linked Nano fiber Composite Anion Exchange Membranes. *Macromolecules* **2014**, *47*, 227–235. [CrossRef]
45. Park, A.M.; Wycisk, R.J.; Ren, X.; Turley, F.E.; Pintauro, P.N. Crosslinked poly(phenylene oxide)-based nanofiber composite membranes for alkaline fuel cells. *J. Mater. Chem. A* **2016**, *4*, 132–141. [CrossRef]
46. Yarin, A.L.; Kataphinan, W.; Reneker, D.H. Branching in electrospinning of nanofibers. *J. Appl. Phys.* **2005**, *98*, 64501. [CrossRef]
47. Tan, S.; Huang, X.; Wu, B. Some fascinating phenomena in electrospinning processes and applications of electrospun nanofibers. *Polym. Int.* **2007**, *56*, 1330–1339. [CrossRef]
48. Reneker, D.H.; Yarin, A.L. Electrospinning jets and polymer nanofibers. *Polymer* **2008**, *49*, 2387–2425. [CrossRef]
49. Gevorkyan, A.; Shter, G.E.; Shmueli, Y.; Buk, A.; Meir, R.; Grader, G.S. Branching effect and morphology control in electrospun PbZr 0.52 Ti 0.48 O 3 nanofibers. *J. Mater. Res.* **2014**, *29*, 1721–1729. [CrossRef]
50. Gottesfeld, S.; Dekel, D.R.; Page, M.; Bae, C.; Yan, Y.; Zelenay, P.; Kim, Y.S. Anion exchange membrane fuel cells: Current status and remaining challenges. *J. Power Sources* **2018**, *375*, 170–184. [CrossRef]
51. Dekel, D.R.; Rasin, I.G.; Brandon, S. Predicting performance stability in anion exchange membrane fuel cells. *J. Power Sources* **2019**, *420*, 118–123. [CrossRef]
52. Omasta, T.J.; Park, A.M.; LaManna, J.M.; Zhang, Y.; Peng, X.; Wang, L.; Jacobson, D.L.; Varcoe, J.R.; Hussey, D.S.; Pivovar, B.S. Beyond catalysis and membranes: Visualizing and solving the challenge of electrode water accumulation and flooding in AEMFCs. *Energy Environ. Sci.* **2018**, *11*, 551–558. [CrossRef]
53. Miller, H.A.; Lavacchi, A.; Vizza, F.; Marelli, M.; Di Benedetto, F.; D'Acapito, F.; Paska, Y.; Page, M.; Dekel, D.R. A Pd/C-CeO $_2$ Anode Catalyst for High-Performance Platinum-Free Anion Exchange Membrane Fuel Cells. *Angew. Chem.* **2016**, *128*, 6108–6111. [CrossRef]
54. Dekel, D.R. Alkaline membrane fuel cell (AMFC) materials and system improvement-state-of-the-art. *ECS Trans.* **2013**, *50*, 2051–2052. [CrossRef]
55. Zheng, Y.; Ash, U.; Pandey, R.P.; Ozioko, A.G.; Ponce-González, J.; Handl, M.; Weissbach, T.; Varcoe, J.R.; Holdcroft, S.; Liberatore, M.W.; et al. Water Uptake Study of Anion Exchange Membranes. *Macromolecules* **2018**, *51*, 3264–3278. [CrossRef]
56. Noga, Z.; Mondal, A.N.; Weissbach, T.; Steven, H.; Dekel, D.R. Effect of CO2 on the properties of anion exchange membranes for fuel cell applications. *J. Member. Sci.* **2019**, *586*, 140–150.
57. Ziv, N.; Mustain, W.E.; Dekel, D.R. The Effect of Ambient Carbon Dioxide on Anion-Exchange Membrane Fuel Cells. *ChemSusChem* **2018**, *11*, 1136–1150. [CrossRef] [PubMed]
58. Zhegur, A.; Gjineci, N.; Willdorf-Cohen, S.; Mondal, A.N.; Diesendruck, C.E.; Gavish, N.; Dekel, D.R. Changes of Anion Exchange Membrane Properties During Chemical Degradation. *ACS Appl. Polym. Mater.* **2020**, *2*, 360–367. [CrossRef]
59. Li, Y.S.; Zhao, T.S.; Yang, W.W. Measurements of water uptake and transport properties in anion-exchange membranes. *Int. J. Hydrog. Energy* **2010**, *35*, 5656–5665. [CrossRef]
60. Kusoglu, A.; Kwong, A.; Clark, K.T.; Gunterman, H.P.; Weber, A.Z. Water Uptake of Fuel-Cell Catalyst Layers. *J. Electrochem. Soc.* **2012**, *159*, F530–F535. [CrossRef]
61. Ziv, N.; Dekel, D.R. A practical method for measuring the true hydroxide conductivity of anion exchange membranes. *Electrochem. Commun.* **2018**, *88*, 109–113. [CrossRef]
62. Cooper, K. Characterizing through-plane and in-plane ionic conductivity of polymer electrolyte membranes. *ECS Trans.* **2011**, *41*, 1371–1380.
63. Ding, C.; Fang, H.; Duan, G.; Zou, Y.; Chen, S.; Hou, H. Investigating the draw ratio and velocity of an electrically charged liquid jet during electrospinning. *RSC Adv.* **2019**, *9*, 13608–13613. [CrossRef]
64. Mann-Lahav, M.; Halabi, M.; Shter, G.E.; Beilin, V.; Balaish, M.; Ein-Eli, Y.; Dekel, D.R.; Grader, G.S. Electrospun ionomeric fibers with anion conducting properties. *Adv. Funct. Mater.* **2019**, 1901733. [CrossRef]

65. Elishav, O.; Grader, G.S. Electrospun Fe-Al-O Nanobelts for Selective CO_2 Hydrogenation to Light Olefins. Unpublished work. 2020.
66. De Vrieze, S.; Van Camp, T.; Nelvig, A.; Hagström, B.; Westbroek, P.; De Clerck, K. The effect of temperature and humidity on electrospinning. *J. Mater. Sci.* **2009**, *44*, 1357. [CrossRef]
67. Pelipenko, J.; Kristl, J.; Janković, B.; Baumgartner, S.; Kocbek, P. The impact of relative humidity during electrospinning on the morphology and mechanical properties of nanofibers. *Int. J. Pharm.* **2013**, *456*, 125–134. [CrossRef] [PubMed]
68. Baumgarten, P.K. Electrostatic spinning of acrylic microfibers. *J. Colloid Interface Sci.* **1971**, *36*, 71–79. [CrossRef]

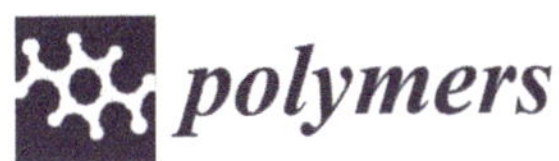

Review

Recent Progress in the Development of Aromatic Polymer-Based Proton Exchange Membranes for Fuel Cell Applications

Raja Rafidah R. S. [1], **Rashmi W.** [2,*], **Khalid M.** [3,*], **Wong W. Y.** [4,*] **and Priyanka J.** [3]

1 School of Engineering, Taylor's University, Subang Jaya 47500, Malaysia; rafeedahrosset@gmail.com
2 Department of Chemical Engineering, School of Energy and Chemical Engineering, Xiamen University Malaysia, Sepang 43900, Malaysia
3 Graphene and Advanced 2D Materials Research Group (GAMRG), School of Science and Technology, Sunway University, Subang Jaya 47500, Malaysia; priyankaj@sunway.edu.my
4 Fuel Cell Institute, UniversitiKebangsaan Malaysia, UKM Bangi, Selangor 43600, Malaysia
* Correspondence: rashmi.walvekar@gmail.com (R.W.); khalids@sunway.edu.my (K.M.); waiyin.wong@ukm.edu.my (W.W.Y.)

Received: 25 February 2020; Accepted: 20 March 2020; Published: 6 May 2020

Abstract: Proton exchange membranes (PEMs) play a pivotal role in fuel cells; conducting protons from the anode to the cathode within the cell's membrane electrode assembles (MEA) separates the reactant fuels and prevents electrons from passing through. High proton conductivity is the most important characteristic of the PEM, as this contributes to the performance and efficiency of the fuel cell. However, it is also important to take into account the membrane's durability to ensure that it canmaintain itsperformance under the actual fuel cell's operating conditions and serve a long lifetime. The current state-of-the-art Nafion membranes are limited due to their high cost, loss of conductivity at elevated temperatures due to dehydration, and fuel crossover. Alternatives to Nafion have become a well-researched topic in recent years. Aromatic-based membranes where the polymer chains are linked together by aromatic rings, alongside varying numbers of ether, ketone, or sulfone functionalities, imide, or benzimidazoles in their structures, are one of the alternatives that show great potential as PEMs due totheir electrochemical, mechanical, and thermal strengths. Membranes based on these polymers, such as poly(aryl ether ketones) (PAEKs) and polyimides (PIs), however, lack a sufficient level of proton conductivity and durability to be practical for use in fuel cells. Therefore, membrane modifications are necessary to overcome their drawbacks. This paper reviews the challenges associated with different types of aromatic-based PEMs, plus the recent approaches that have been adopted to enhance their properties and performance.

Keywords: fuel cells; aromatic-based; polymers; proton exchange membranes; modifications

---

1. Introduction

A solid ion-conducting electrolyte membrane is one of the vital core components in fuel cell systems, namely for the types operating at a temperature range between room temperature and 200 °C, such as Low and High Temperature Proton Exchange Membrane Fuel Cells (LTPEMFCs and HTPEMFCs), Anion Exchange Membrane Fuel Cell (AEMFCs), Direct Methanol Fuel Cell (DMFCs), and Microbial Fuel Cell (MFCs) [1]. The solid electrolyte functions as a separator between the anode and cathode, repels electrons and acts as a barrier between thefuel and oxidants [2,3]. Proton exchange membranes (PEMs) applied in PEMFCs and DMFCs are cationic exchange membranes possessing negatively charged groups (SO_3^-, -COO$^-$, -PO_3^{2-}, etc.) on the membrane's polymeric backbone that provide a conducting pathway for cations, normally protons, but reject anions. Conversely, the polymeric backbone of anion exchange membranes (AEMs) for AEMFCs holdspositively charged groups (-NH_2^+, -NR_2H^+, -PR^+, etc.) for the transport of anions, such as hydroxides [4]. Achieving high performance in fuel cells requires that these ion exchange membranes have desired characteristics in terms of their electrochemical properties and durability, such asa high proton/anion conductivity (approximately, σ = 0.1 S/cm); low electron conductivity; good resistance to fuel crossover; excellent mechanical, thermal, and chemical strength; acceptable hydrolytic and oxidative stability; and low cost of fabrication and assembly [4–8].

Perfluorosulfonic acid (PFSA)-based polymers, namely the Nafion membrane, have been the standard for fuel cell PEMs due to their high proton conductivity and excellent durability. However, the performance of Nafion is affected by its poor methanol resistance and drop in proton conductivity under high temperatures and low humidity conditions. Additionally, the membrane is also expensive [9,10]. In fuel cell production, the cost of the stack covers 66% of the whole system, and the membrane contributes to 17% of the stack cost for the production of 1000 systems per year (DOE Fuel Cell Report 2017) [11]. Efforts to reduce the cost and improve the performance of each component in the fuel cell stack have been made by researchers in recent years. For membranes, several potential, low-cost alternatives to Nafionhavealready been extensively studied. Among these alternatives to the Nafion membrane are biopolymers based on chitosan, alginates, or cellulose, and the non-fluorinated or partially-fluorinated hydrocarbon polymer membranes with aromatic backbone structures. While biopolymers have advantages in terms of their renewability, durable properties under fuel cell operating conditions, and lower cost, aromatic-based membranes are equally as advantageous, having an excellent thermal and mechanical strength, tailorable structures, tunable ionic conductivities, smaller methanol permeabilities, and potentially lower costs. Past reviews on alternative PEMs have highlighted the different properties, modifications, and performances ofseveral types of aromatic-based membranes alongside Nafion and biopolymers. However, few recent reviews have focused solely on aromatic-based membranes. This paper aims to discuss the challenges associated with several known aromatic-based polymers in applications to LTPEMFCs, HTPEMFCs, and DMFCs, which include highlights of the latest studies on research related to their modifications, improvements, durabilities, and performances as protonicexchange membranes.

2. Types of Aromatic Polymer-Based PEMs: Properties and Development

The backbones of these polymers are linked together by aromatic and phenyl rings with C-C, C=C, and C-H bonds within their backbones that provide the membrane with excellent mechanical, thermal, and chemical strengths. These linkages include groups with varying numbers of ether and ketone functionalities, such as those in poly(ether ketone)s (PEKs) and poly(aryl ether ketone)s (PAEKs)-type polymers; sulfone functionalities in poly(ether sulfone)s (PESs), and polysulfones (PSFs); imide bonds in polyimides (PIs); benzimidazole rings in polybenzimidazole(PBI)-based polymers;ether-containing polyphenylene oxide (PPO), and more. Typical unit structures of the mentioned polymers are shown in Table 1. he ionic conducting properties of these membranes are ineffective in theirpristine form. To cater for their conductive properties, strong acidic, proton conductive sulfonic acid groups (-SO_3H) are commonly introduced into these polymer chains through reactions with sulfonating agents, producing sulfonated derivatives of these polymers that are more suitable for application as PEMs. Other functionalities are also possible, depending on the final application, whether as PEMs for low- or high-temperature PEMFCs, or as AEMs. Examples of these other functionalities are quaternary ammonium, imidazolium, or benzimidazole groups [4,12–16].

Past research has highlighted that the strength of the proton conductivity of these membranes is governed by the concentration of effective ionic/proton conductive groups (referred to as the ion exchange capacity (IEC)), hydration levels, temperature, and hydrophilic/hydrophobic phase separation. Typically, within hydrated membranes, proton conduction occurs through the Grotthuss mechanism (proton hopping between ionic domains and water molecules) and vehicle mechanism (proton diffusion). In most cases for sulfonated polymers, the degree of sulfonation (DS) or IEC determines their hydrophilicity and conductivity, where a higher DS/IEC leads to a more hydrophilic membrane with better water uptake, hydration, and conductivities. However, a highly hydrophilic membrane tends to absorb water excessively. Weakened interactions between polymer chains through ionic and hydrogen bonds resulting from a large number of water molecules occupying the free volumes between chains can cause a large amount of swelling, mechanical deterioration, and other issues related to the oxidative stability and reactant permeabilities [15,17]. Furthermore, the smaller hydrophobic/hydrophilic phase separation of these polymers in comparison to Nafion leads to lowered methanol permeabilities, providing advantages in terms of the performance in methanol fuel cells [18,19]. However, researchers have also noted that the small phase separation contributes to low conductivities due to less connected ionic domains, meaning the formation of proton conducting channels is not as effective as in Nafion [20,21].

The mentioned advantages and disadvantages are shared bymost of the aromatic-based membranes considered in this review. However, other specific strengths and weaknesses may be displayed by individual polymers. Understanding the effects of the characteristic at a molecular level towards the function of the PEM as a whole candetermine their electrochemical performance, their durability under varying fuel cell conditions, and their lifetime. Past researchers have adopted strategies and modification techniques to design hydrocarbon-based PEMs with improved properties and performances. To utilize the advantage of high proton conductivities offered by these PEMs with high DS/IEC requires balance with their water uptake and mechanical strength, which remains a challenging task. Overthe years, several of the methods that have been appliedby researchers in an attempt to improve the properties of hydrocarbon-based PEMs include structural modifications (additional branching, pendant groups, etc.), crosslinking, polymer blending, mixed-matrix, block copolymerization, and the introduction of inorganic/organic fillers/nanofillers [5,7,22–24]. While these modifications yielded positive enhancements ofthe PEM, it is important to consider optimization between electrochemical properties; thermal, mechanical, and chemical strengths; water uptake; fuel resistance; and oxidative stability by carefully designing the polymeric structure andcontrolling the ratio of combination between materials. The potential to effectively utilize these PEMs in actual fuel cell conditions depends on how well their individual characteristics balance out.

Table 1. Typical unit structures of several aromatic-based polymers used as proton exchange membrane (PEM) materials.

Polymer Name	**Typical Unit Structures**
Poly aryl ether ketone (PAEK)	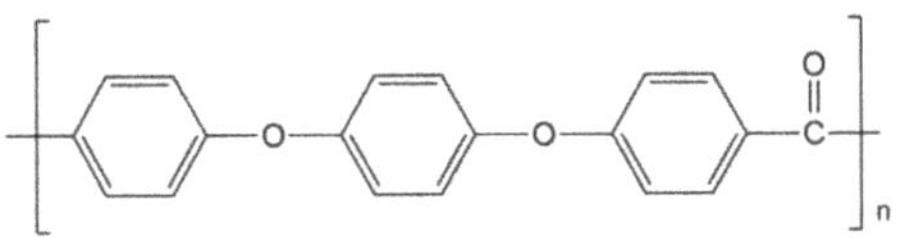 (a) Polyether ether ketone (PEEK) 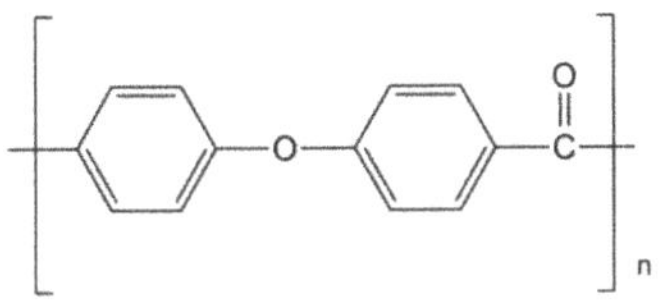 (b) Polyether ketone (PEK)

Table 1. *Cont.*

Polymer Name	Typical Unit Structures
Polyimide (PI)	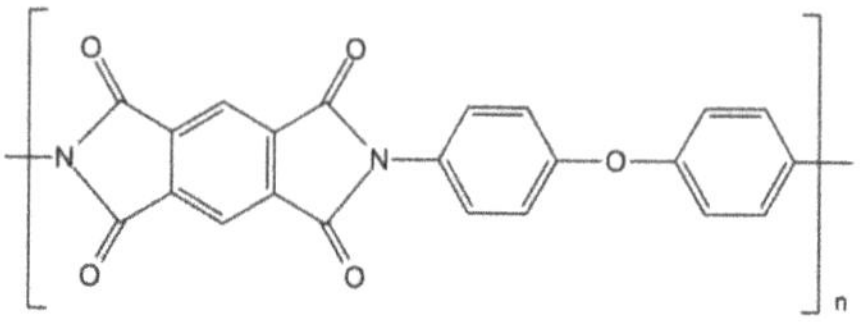 (a) Aromatic polyimide
Polysulfone (PSF or PSU)	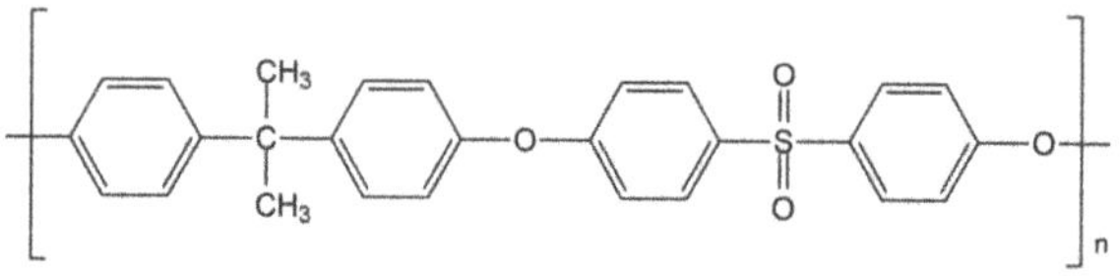 (a) PSF with alkyl linkage 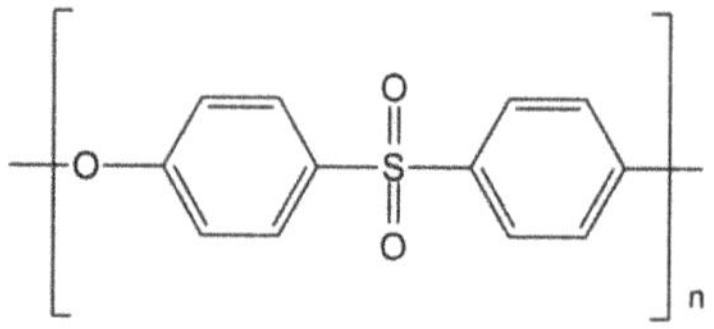 (b) Polyether sulfone (PES)
Polybenzimidazole (PBI)	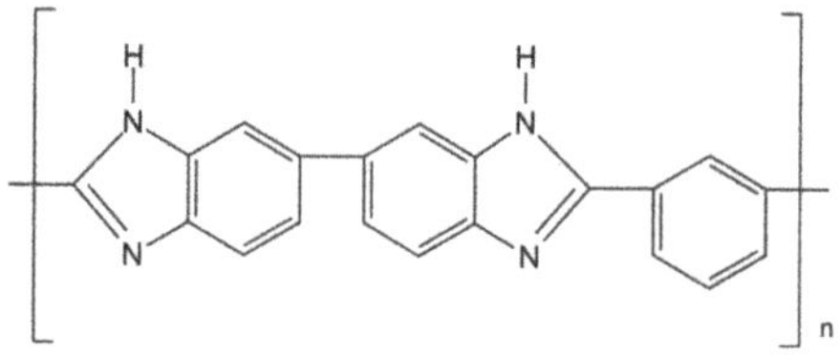 (a) meta(m)-PBI 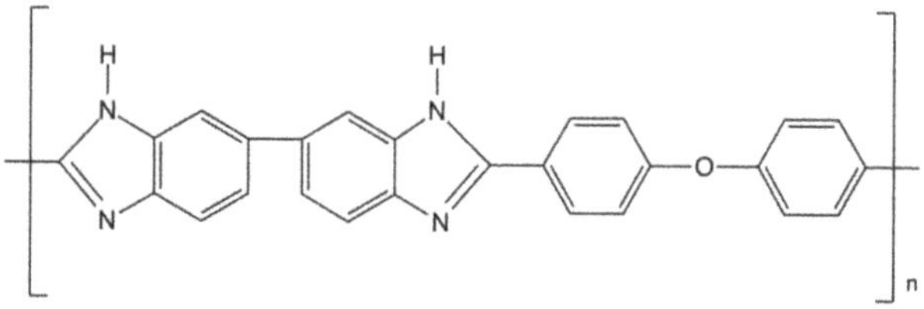 (b) PBI with ether linkage (OPBI)
Polyphenylene oxide (PPO)	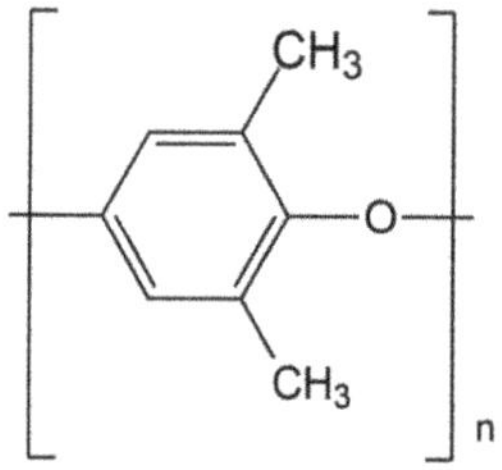

2.1. Poly Aryl Ether Ketones (PAEKs)

Poly aryl ether ketones (PAEKs) refer to polymers consisting of different numbers of ether and ketone functionalities connecting aryl rings, such as polyether ketones (PEKs), polyether ether ketones (PEEKs), and polyether ketone ketones (PEKKs). More ether or ketone groups may be included and their positions within the chain depend on the monomers used at the beginning of PAEK synthesis. The chain may also contain alkyl groups or fluorinated functional groups. These polymers are semicrystalline thermoplastics withgood chemical and thermal stability, dielectric properties, and mechanical strength. Sulfonated poly(ether ether ketone) (SPEEK) is -SO_3H-functionalized PEEK andis one of the most commonly studied PEMs for fuel cells. Its synthesis is normally carried out through the post-sulfonation of commercial PEEK with strong sulfuric acid (98% H_2SO_4). K. Kreuer [25] has proposed the difference between the microstructural arrangement of –SO_3H groups in Nafion and the two-ketone SPEEKK, where there is less pronounced hydrophobic/hydrophilic phase separation in SPEEKK compared to Nafion. This leads to poorer connectivity between –SO_3H ionic domains, and thus a smaller proton conductivity of SPEEKK. PEEK with cationic groups, such as quaternary ammonium and imidazolium, may be utilized as AEM [16].

Recent studies have focused on understanding the PEM properties of SPEEK membranes, including durability studies, which are important in predicting the membrane's lifetime under fuel cell operating conditions [17,26–28]. Regarding the pure SPEEK, M. Parnian et al. [17] highlighted the changes in several key properties of SPEEK in their study of SPEEK with various DS values. The summary of some of their findings, stated in Table 2, revealed an increasing trend in water uptake, swelling, proton conductivity, and thermal degradation at an increasing DS, but deterioration in the tensile and oxidative stability. The oxidative stability refers to the membrane's resistance to degradation due to peroxide radicals (HO• and HOO•) formed from incomplete oxygen reduction reactions at the fuel cell cathode. The SPEEK with the lowest %DS takes the longest time to completely disintegrate in Fenton's reagent. With increasing %DS and higher water uptake, the radicals are more easily diffused into the membrane and attack the aromatic backbone, sothe membrane becomes rapidly disintegrated.

Table 2. Property changes in the sulfonated poly(ether ether ketone) (SPEEK) membrane with varying degrees of sulfonation [17].

Degree of Sulfonation (%)	Water Uptake (%, RT)	Thickness Swelling Ratio (%, RT)	Proton Conductivity (S/cm, 80 °C)	Tensile Strength (MPa)	Thermal Stability (% Degradation to 600 °C)	Oxidative Stability (~min)
40.23	6.29	2.13	0.2571	73	44	200
65.52	14.62	12.44	0.3003	63	46	56
75.95	52.01	27.20	0.4252	50.45	50	<6
89.23	97.98	34.54	0.4649	41	56	<2

In terms of the processability and MEA performance, common organic solvents, namely DMSO, DMAc, DMF, and NMP, used for the solution casting of SPEEK also seemto influence the conductivity of high and low DS SPEEK. X. Liu et al. [29] found that low DS SPEEK casted from DMSO has a larger conductivity due to the weaker molecular interactions between residual solvent molecules and the polymer, while no significant change in properties wasobserved, regardless of the solvent type, for high DS SPEEK. The ex-situ mechanical degradation of SPEEK in hygrothermal cycle tests for 700 min conducted by S. H. Mirfasi et al. [27] showed permanent deformation and a 4 micron reduction of membrane thickness at the end of the tests, resulting in a faster hydrogen crossover rate. Under creep and tensile residual stress, the membrane wasdegraded due to fatigue and became more brittle, and the toughness dropped. Additionally, the degraded SPEEK exhibited an increase in water uptake and swelling-induced stress, thusworsening the dimensional stability. A. Karimi et al. [28] provided a study on the MEA model of SPEEK compared to Nafion MEA. The use of a SPEEK membrane in MEA at high temperatures is favorable when the pressure and water content at the anode gas feed arehigher, but the cell performance drops when the water activity is largely reduced at 140 °C. Additionally, the SPEEK proton conduction is smaller than that of Nafion.

Improvements to SPEEK and Other PAEK-Based PEMs

Recently, researchers followed similar strategies to the modifications of SPEEK with the use of new or modified materials to improve their properties. Reaping the advantages of high DS SPEEK requires the intermolecular interactions between polymeric chains to bestrong enough to overcome swelling, especially at high temperatures. S. Gao et al. [30] investigated the properties of nanocomposite SPEEK with a high DS of 84% grafted with graphene oxide (GO-g-SPEEK), synthesized from the 'grafting' reaction between partially hydroxyl-functionalized SPEEK and brominated GO (GO-Br). The restraining effect of GO limited the SPEEK's swelling, despite thetriple increment in water uptake compared to Nafion. Furthermore, its conductivity improved above 80 °C, which is the point where the proton conductivity of Nafion would drop due to dehydration. Blends of GO-g-SPEEK with Nafion-33 achieved a conductivity of around 0.22 S/cm at 90 °C, while its MEA performance reached a power density of 213 mW/cm^2 compared to 112 mW/cm^2 for unblended GO-g-SPEEK. This suggested that blending with Nafion provided better interfacial contact between the catalyst and membrane. Another study by S. Bano et al. [31] focused on crosslinked SPEEK with ethylene glycol (EG), followed by the incorporation of cellulose nanocrystals (CNCs). XRD analysis of the resulting XSPEEK-CNC showed the amorphous characteristic of the membranes. The hydrophilic sulfate and hydroxyl groups present on CNCs restore the loss in water uptake and IEC due to the crosslinking of SPEEK. An effective increment in proton conductivity is observed up to 4 wt% of CNC, attributed to the homogenous dispersion of CNCs, the presence of additional hydrophilic functionalities, interfacial hydrogen bonds, and good hydrophilic-hydrophobic phase separation. However, the XSPEEK-CNC is also a stiff membrane as a result of crosslinking.

Morphological modifications through the introduction of nanofibers also showed several advantages. G. Liu et al. [32] studied the properties of SPEEK-impregnated poly(vinylidene fluoride) (PVDF) electrospun nanofibers for DMFC. SPEEK (76.7% DS) wasembedded in PVDF nanofibers through a cloud point polymer/solvent/non-solvent (SPEEK/DMAc/H_2O) ternary system (Figure 1) to prevent PVDF nanofibers from collapsing when mixed with SPEEK (as PVDF is also soluble in DMAc).The SPEEK-embedded PVDF nanofiber membrane wasfound to be more flexible with a sustained yield and modulus. Its methanol permeability washalf of that of pristine SPEEK and 1/3 of that of Nafion 115. This advantageous property led to a higher peak power density at 104 mW/cm^2 compared to Nafion 115 at 84 mW/cm^2 in DMFC MEA that utilized 5M methanol. Y. Wu et al. [33] fabricated SPEEK core-shell nanofibers containing sulfonated organosilane graphene oxide (SSi-GO) nanosheets through coaxial electrospinning, forming a cambiform-like morphology in the membrane. The co-spinning SSi-GO/SPEEK exhibited higher water uptake than the membranes formed from casting and mono-spinning, where wrinkled voids in SSi-GO weresuggested to act as a water reservoir, while the SSi-GO restricted swelling. The methanol permeability was 39% of that of Nafion 115, despite being thinner. The proton conductivity of the membrane containing 2.5 wt% SSi-GO was1.24 and 1.42 times higher than that of mono-spinning and casted membranes, respectively, likely due to the uniform dispersion of SSi-GO, axial arrangement of SSi-GO with SPEEK nanofibers to form cambiform core-shell nanofibers (Figure 2), higher water uptake, and conductive sulfonic and hydroxyl groups.

In terms of other PAEK membranes, K. Kang et al. [34] investigated the properties of a semi-interpenetrating network (IPN) consisting of a multiblock copolymer of PAEK-b-KSPAEK with organosiloxane (OSPN). The OSPN-containing membrane displayed an increased elasticity and reduced permeation of oxidative radicals and methanol. Because of the hydrophilic characteristics of OSPN, which elevated the PAEK-b-KSPAEK's water absorption, the proton conductivity achieved was close to that ofNafion 115. PAEK polymers have also been utilized in HTPEMFC. J. Li et al. [35] characterized the properties of a PA-doped brominated tetramethyl PAEK (BrPAEK) decorated with four types of nitrogen-heterocyclic molecules (Pyridine, 1-methylimidazole, 1H-benzotriazole, and 3-amino-1,2,4-triazole). Pyridine- and 1-methylimidazole-containing BrPAEK were the only membranes showing a PA absorption ability, andthe latter contributed to the BrPAEK with the highest conductivity at 170 °C, in ananhydrous state. The conductivity exceeded that of PBI when there was asufficient content of imidazole per unit, in which a larger PA doping level waspossible. However, more imidazole per unit negatively affects the oxidaticve stability. Furthermore, J. Yang et al. [36] studied the effects of blending PVDF and PVDF-HP (poly(vinylidene fluoride-co-hexafluoride) with 1-methylimidazole-containing BrPAEK (Melm-PAEK). The presence of PVDF and PVDF-HFP stabilized the membrane by reducing the PA uptake and swelling, as well as enhancing the tensile strength, as the pristine Melm-PAEK was unstable due to the high bromination degree (45%). Blends with PVDF-HP exhibited better PA uptake and proton conductivity, yet a smaller mechanical strength, which was likely due to the presence of large trifluoromethyl groups.

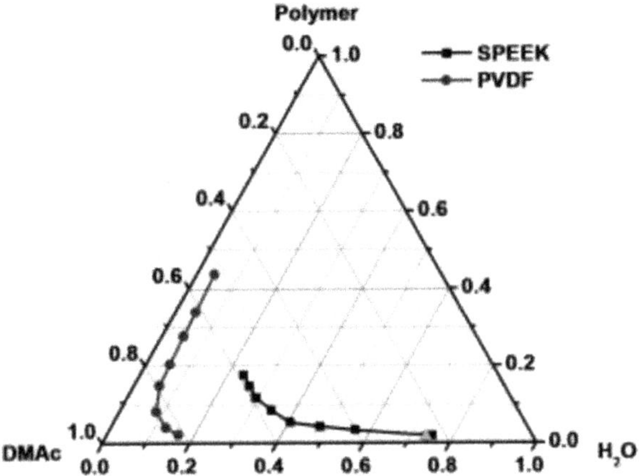

Figure 1. Cloud point curves of SPEEK/dimethylacetamide (DMAc)/H_2O (square plot) and poly(vinylidene fluoride) (PVDF)/DMAc/H_2O (circle plot) [32].

Table 3 summarizes some of the recent literature on the improvements to PAEK and PEEK membranes. Focus is given tothe proton conductivity, as this influences the PEM performance, along with IEC and water uptake, which affects proton transport. The electrochemical aspects do not show exhibitdifferences between PAEK and PEEK, and upon functionalization to SPEEK and SPAEK or other functionalities, they show a similar dependency to DS, IEC, water uptake, and PA uptake (HT-PEMFC application). Their thermal strengths arewithin the range of fuel cell operations below 200°C. Furthermore, their mechanical stabilities also depend on the aforementioned factors, where modifications that enhance the strong intermolecular interactions between polymer chains, such as crosslinking and the use of fillers, can allow the utilization of SPEEK or SPAEK with a higher DS. It is noted that more literature seems to be available for SPEEK compared to other SPAEK- or PAEK-type polymers. SPEEK's synthesis through the post-sulfonation of commercially available PEEK is likely simpler and the DS is easier to control, while more complex polymerization reactions may be required to prepare other PAEK-type polymers with differing structures.

Table 3. Water uptake and electrochemical properties of PAEK-type PEMs.

PAEK-Type									
Membrane	**Year**	**%DS**	**Modifications**	**Fuel Cell Type**	**Filler Content**	**IEC (meq/g)**	**Water Uptake (%)**	**Proton Conductivity (S/cm)**	**Peak Power Density (mW/cm^2)**
SPEEK/n-BuOH [37]	2015	-	n-BuOH self-organization inducer	PEMFC	-	1.5 (mmol/g)	~52 (80°C)	0.314 (80°C)	-
b-CPAEK [38]	2016	-	PAEK block copolymers	DMFC	-	1.92	~50 (90°C)	~0.11 (90°C, 95%RH)	-
SPEEK/AIT [39]	2016	68	Amine-functionalized iron titanate (AIT)	PEMFC	2wt% AIT	-	72 (25°C)	0.12 (80°C)	204 (80°C, 90%RH, H_2/O_2)
Pore filling SPAEK [40]	2017	80	SPAEK-filled porous PAEK	DMFC	-	1.47	~55 (90°C)	~0.11 (90°C, 90% RH)	-
SPEEK-SrGO [41]	2017	-	Sulfonated reduced graphene oxide	PEMFC	1wt% SrGO	1.69	31.1 (80°C)	0.0086 (80°C 50%RH)	705 (70°C,80%RH, H_2/Air)
BrPAEK-MeIm [35]	2018	-	Nitrogen-heterocycles	HTPEMFC	1.6 imidazole/unit	1.95	-	0.091 (170°C, 0% RH)	-
MeIm-PAEK/PVDF-HFP [36]	2018	-	MeIm-PAEK/PVDF-HFP blend	HTPEMFC	10% PVDF-HFP	-	103 (60°C)	0.219 (180°C, 0%RH)	-
SPEEK/Bu/SPEEK/Im [42]	2018	47	SPEEK/PU/SPEEK/bmim layer-by-layer	HTPEMFC	-	-	-	0.103 (160°C, 0%RH)	-
GO-g-SPEEK/Nafion-33 [30]	2018	80	GO, Nafion-33 blended	PEMFC	-	1.45	136.3 (90°C)	~0.23 (90°C)	213 (60°C, 50% RH, H_2/Air)
XSPEEK/CNC [31]	2019	70	EG + CNC	PEMFC	67:33 (SPEEK:EG) 4wt% CNC	1.72	78.2 (95°C)	0.186 (95°C,95%RH)	-
SPEEK/PDA@PVDF [32]	2019	76.7	SPEEK embedded PDA-containing PVDF nanofibers	DMFC	85.7 wt% PDA@PVDF	-	~60 (80°C)	0.06 (80°C, 100%RH)	104 (5 M MeOH/O_2, 70°C)
SSi-GO/SPEEK [33]	2019	-	SPEEK nanofibers/SSi-GO	DMFC	2.5wt%	1.65 (mmol/g)	~90 (70°C)	0.1566 (70°C,100% RH)	-
PAEK-b-KSPAEK/OSPN [34]	2019	-	PAEK-b-KSPAEK copolymer/OSPN	PEMFC	24wt% OSPN	-	84.01 (90°C)	~0.11 (90°C, 100%RH)	410 (80°C, 100%RH, H_2/O_2)

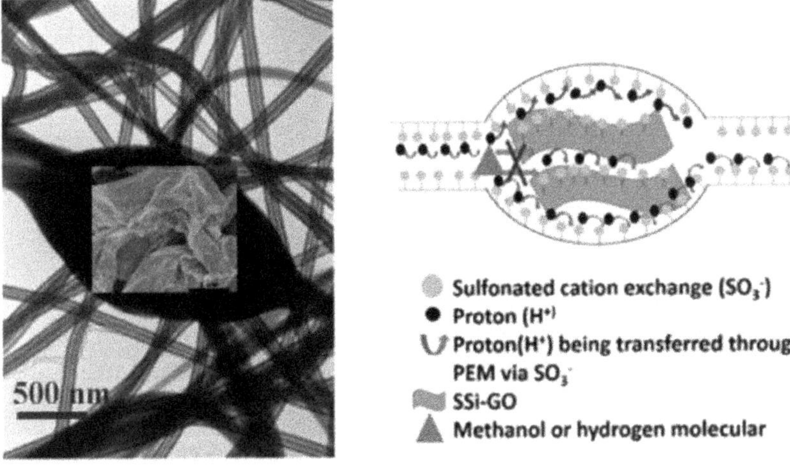

Figure 2. TEM image of a co-spinning sulfonated organosilane graphene oxide (SSi-GO)/SPEEK and aschematic representation of methanol and hydrogen diffusion through the cambiform-like structure of the core-shell nanofibers [33].

2.2. Polyimide (PI)

The formation of charge transfer interactions between the dianhydride and diamine functionalities of the polyimide (PI) backbone provides the polymer withexcellent thermal stability. PI membranes also exhibithighmechanical strength and chemical stability. The structure and properties are tailorable by using different dianhydride and diamine monomers. Sulfonic acid-containing sulfonated polyimides (SPIs) synthesized by using sulfonated precursors, such as 4,4′-diaminodiphenyl ether-2,2′-disulfonic acid (ODADS), 4,4-diaminostilbene-2,2-disulfonic acid (DSDSA), and 4,4′-Diamino-2,2′-biphenyldisulfonic (DAPS), result in PI membranes with proton conductive properties [12,43,44] Final structures of PIs are tailorable and block copolymers may also be formed.Aside from thechallenges related to ahigh DS and IEC, it has also been highlighted that the imide rings are very sensitive to water, sothe unit is prone to hydrolysis, negatively impacting the hydrolytic stability [45].

Improvements to SPI PEMs

In recent years, several studies have investigated the potential of SPIs with varying molecular structures, while polymer blending and filler additives have also been attempted. Z. Yao et al. [46] aimed to improve the hydrolytic stability of SPIs by synthesizing perylenediimine-containing, aliphatic-type SPIs with different chain lengths, through the mild polyacylation of a sulfonated diarene monomer and aliphatic perylenediimide dicarboxylic acid monomers. Both short and long aliphatic SPIs showedimproved hydrolytic stability, compared to normal SPIs, under tests in hot water at 80 °C for 300 h, where the likely factors included the rigid perylene structures that helped strengthen the polymer. However, the long-chain AL-SPI-10 became a bit fragile after the test. Furthermore, shorter-chain AL-SPI-5exhibited better IEC and water uptake than the long-chain counterpart, while its proton conductivity was higher than that ofNafion 115 at 80 °C. K. Liaqat et al. [47] characterized a novel sulfonated polyimide (NSPI) with a unique structure inwhich the $-SO_3H$ groups wereattached to a phenylene side chain rather than the main chain, via the copolymerization of novel sulfonated diamine (NSDA) with 4,4′-oxydianiline (ODA) and 4,4′-oxydiphthalic anhydride (ODPA). The relocation of $-SO_3H$ groups to the side chain was suggested to improve the hydrolytic stability of NSPI under the test at 140 °C with pressurized steam, by lowering the risk of the nucleophilic attack of hydroxyl ions on the imide rings. The power density of the NSPI with a 10/90 ratio of NSDA/ODA exhibited a slightly elevated peak power density of 28.7 mW/cm^2 compared to Nafion 117 (25 mW/cm^2) in DMFC MEA at 70 °C. These reports highlight that a better hydrolytic and oxidative stability of SPI ispossible if there is a distinct separation between the –SO_3H groups andthe imide rings.

S. Feng et al. [48] applied the charge transfer (CT) complex method to prepare CT complex blend membranes using electron-accepting SPIs and electron-donating polyethers (PEs). The SPI/PE CT complexes were subjected to heat treatment. While the heat-treated membranes displayed improvements in thetensile strength, elongation at break (EB), and Young's modulus compared to non-heat-treated membranes and normal SPI, other properties were characterized for non-heat-treated membranes due to issues with heat treatment and reproducibility. The thin CT complex membranes with a thickness from 13 to 28 microns displayed hydrogen permeabilities that were 4.1 and 5.4 times smaller than those of Nafion 212, which wasconsidered as an advantageous characteristic for thin membranes. Despite the smaller values of IEC, proton conductivity, and peak power density, the thin SPI/PE was found to be stable under the 10 h MEA test. Nanofibers based on different structures of SPIs were investigated by G. Ito et al. [49]. Block-types SPIs and graft-type SPIs both exhibited the properties of self-assembling hydrophilic/hydrophobic microphase structures in a membrane state. It was found that the fluorine-concentrated SPI nanofibers hadlower proton conductivities that were attributed to their fragility when dehydrated. On the other hand, less-fluorinated sulfonated random polyimide (S-r-PI) nanofibers achieved mechanically stable, low gas permeation, and high proton conductive characteristics. Hence, the S-r-PI nanofibers were integrated into a sulfonated block-graft polyimide (S-bg-PI) polymer matrix to form a Nanofiber Framework (NfF) composite membrane with a 12 micron thickness. The nanofibers reinforced the membrane mechanically and improved the thin membrane's gas permeability. Furthermore, the composite's proton conductivity was comparable to and even exceeded that, of Nafion NR212 at temperature range between 30 to 90 °C and 95% RH.

Aside from structural modifications, filler addition to the SPI matrix has also shownseveral improvements. Recently, P. Y. You et al. [50] incorporated rice husk ash (RHA) biofillers into the SPI matrix. A suitable concentration of RHA resulted in stiffer membranes and higher water uptake than normal SPIs. At a 15 wt% RHA content, the proton conductivity was found to be double that of Nafion. This improvement canlikely be attributed tothe Lewis acid-base interaction between SPI chains, and the hydroxyl groups of RHA, which attract water molecules (high water retention). However, it should be expected that alarge amount of RHA can block the continuous proton transfer channels.

Some of the recent studies onSPI membranes are summarized in Table 4. DS is not normally specified for SPI, yet the trend of water uptake and conductivity still follows that of IEC, as reported. SPI membranes are physically stable enough, even under conditions of HTPEMFC, providing that the PEM's water or PA retention, mechanical strengths, and hydrolytic stabilities are controlled.

Table 4. Water uptake and electrochemical properties ofpolyimide(PI)-type PEMs.

PI-type									
Membrane	**Year**	**%DS**	**Modifications**	**Fuel Cell Type**	**Filler Content**	**IEC(meq/g)**	**Water Uptake (%)**	**Proton Conductivity (S/cm)**	**Peak Power Density (mW/cm^2)**
SPI/FGO [43]	2015	-	Ionic liquid-functionalized graphene oxide	HTPEMFC	5wt% FGO	-	47.3	0.0772 (160°C, 40%RH)	-
CSiSPIBI [51]	2016	-	Silane-crosslinked sulfonated poly(imide benzimidazole)	HTPEMFC	60 mol fraction sulfonated diamine monomer	0.54	-	~0.1 (150°C,50%RH)	-
CSPI [44]	2017	-	Crosslinked SPI with pendant alkyl side chains containing trimethoxysilyl	DMFC	70 mol% DAPS groups	2.02 (mmol/g)	73.4 (80°C)	~0.13 (80°C,100%RH)	84.3 (2M MeOH/air, 60°C)
Aliphatic SPI [46]	2018	-	Aliphatic SPI with perylenediimide units	PEMFC	-	1.79 (mmol/g)	80 (80°C)	0.1864 (80°C, 100%RH)	931.88 (80°C, 100% RH, H_2/O_2)
NSPI [47]	2018	-	Novel SPI from NSDA/ODA	DMFC	50/50 (wt NSDA/wt ODA)	1.25	38.21 (35°C)		-
SPI-PE [48]	2018	-	SPI-PE charge transfer complex	PEMFC	0.33 molar ratio PE	2.16 (mmol/g)	45.9 (RT)	0.0201 (80°C, 90%RH)	~150 (80°C, 95%RH, H_2/Air)
SPI Nanofiber framework [49]	2018	-	S-block graft (bg)-PI/S-r-PI nanofibers	PEMFC	80/20 (wt S-bg-PI/wt S-r-PI)	1.8	73.9 (RT)	>0.1 (80°C, 85%RH)	-
SPI-RHA [50]	2019	-	SPI-rice husk ash biofillers	Passive-DMFC	15 wt% RHA	0.2519 (mmol/g)	55.24	0.2058 (RT)	13 (2M MeOH, RT)

2.3. Polyether Sulfones (PESs) and Polysulfones (PSFs)

Polysulfones (PSFs or PSUs) are polymers consisting of sulfone and ether linkages. PSF refers to apolymer that also contains an alkyl group, while the shorter chain polyethersulfone (PES) refers to apolymer chain with only sulfone and ether groups. Similar to PAEK and PI, PSF and PES polymers possess high mechanical, thermal, and chemical strengths. PES is denser, with a higher glass transition temperature (~220 °C) compared to PSF (~185 °C) [52]. A number of research studies are available for AEMs based on PSFs, although PSF membranes are also utilized as PEMs. The synthesis of proton conducting sulfonated polyether sulfone (SPES) may be done through using precursors such as 4,4′-Dichlorodiphenylsulfone (DCDPS), or the sulfonation of commercial SPES [53,54]. Related durability issues, particularly theloss in mechanical strength upon a higher degree of sulfonation, limit this polymer's long-term lifetime under an actual fuel cell environment, requiring necessary modification strategies to tailor the morphological and molecular aspects.

Improvements to PES- and PSF-type PEMs

Improvements inPSF and PES membranes are also focused on increasing theirconductivity and durability. S. Matsushita and J. D. Kim [55] attempted the thermal crosslinking of sulfonated poly(phenylsulfone) (SPPSU) with ethylene glycol and glycerol crosslinkers using water as a solvent, with annealing temperatures ranging from 80 to 200 °C. SPPSU prepared with 6 mol/rpu ethylene glycol at 200 °Cwas the membrane with optimized dimensional properties, with an acceptable proton conductivity, despite being lower than 0.1 S/cm. Nevertheless, this study suggested the possibility of using water instead of organic solvents in membrane preparation, as well as the importance of controlling the membrane's physical stability, as there are concerns that it may be damaged during MEA preparation. N. Urena et al. [56] attempted the "one-pot two-step synthesis" approach to synthesize high molecular weight, multiblock copolymers composed of PSU and PPSU using commercially available monomers, followed by sulfonation using trimethylsilyl chlorosulfate. This approach is suggested to be less complex than the polycondensation of monomers forproducing high molecular weight polymers, which may be better for industrial-scale processes. The obtained sulfonated multiblock copolymers (SPSU/SPPSU) had no obvious phase separation but achieved a higher hydration level than Nafion 117. The tensile strengths of the dry membranes exceeded that of Nafion 112, which increased with IEC. Tensile drop was more prominent for the high IEC membrane in a wet state, but still considered as acceptable (55 MPa for the membrane with 1.58 meq/g IEC). The maximum power density of the multiblock SPSU/SPPSU was ~400 mW/cm^2, which was in between Nafion 112 (729 mW/cm^2) and Nafion 117 (310 mW/cm^2) at 70 °C and 100% RH.

S. Gahlot et al. [57] studied the effects of sulfonated mesoporous silica (S-MCM-41) on SPES. Similar to most inorganic-organic composites with porous, hygroscopic fillers, the presence of S-MCM-41 in the SPES matrix increased thewater uptake and tensile strength. The content of bound water was 0.51% for the filler-containing SPES and 0.2% for normal SPES, showing the filler's ability in water retention. While IEC showed an increasing trend forthe S-MCM-41 content in SPES, the highest proton conductivity was achieved with 2 wt% S-MCM-4, due to the uniform filler distribution, higher porosity, and proper ion channel formation. X. Xu et al. [58] prepared cellulose whiskers (CW) functionalized with various Fmoc-amino acids (Glycine, 5-amino-Valeric acid, L-serine, L-Aspargine, and L-Leucine). The Fmoc-protecting groups were removed and incorporated into the SPSF matrix to form a proton conductive mixed-matrix membrane. SPSF containing 10wt% of L-serine-functionalized CW achieved the highest conductivity at 0.234 S/cm at 80 °C. The highly hydrophilic functionalized CW and presence of $-NH_2$ from amino acids provided the membrane with more water and conductive groups. Moreover, the methanol resistance was improved. The power density of SPSF/CW-Ser projected a peak power density of73.757 mW/cm^2, whereas for Nafion 117, thevalue was 51.323 mW/cm^2, at 60 °C and 100% RH in single-cell DMFC MEA with 2M methanol.

A few recent studies have also characterized modified PES-type polymers as PEMs for HTPEMFCs. N. Anahidzade et al. [59] utilized an amino-functionalized metal organic framework (MOF)

(Cr-MIL-101-NH_2) on chlorosulfonatedpoly(ether sulfone) (PES-SO_2Cl), resulting in crosslinking (Figure 3). While the PA uptake was less, the PA retention, proton conductivity, and thermal and mechanical strengths werebetter for the MOF-containing membrane. The conductivity achieved was 0.041 S/cm at 160°C, in an anhydrous state. Moreover, a 14 day durability test under the same conditions also highlighted a less significant drop in conductivity within two days, before remaining stable until the end of test, showing the advantage of the acid-stable, porous MOF towards PEM stability. H. Bai et al. [60] introduced graphitic carbon nitride (CN) nanosheets into a poly(ether sulfone)-poly(vinyl pyrrolidone) (PES-PVP) matrix. CN nanosheets raised the PA uptake and proton conductivity of the composites to a 0.5 wt% nanosheet content. The proton conductivity increased by 36% for the composite with 0.5 wt% CN, compared to the pure PES-PVP at 160°C, in an anhydrous state. The CN nanosheets weresuggested to interact with PA molecules and improve the rate of ionization of free PA and protons from PVP.

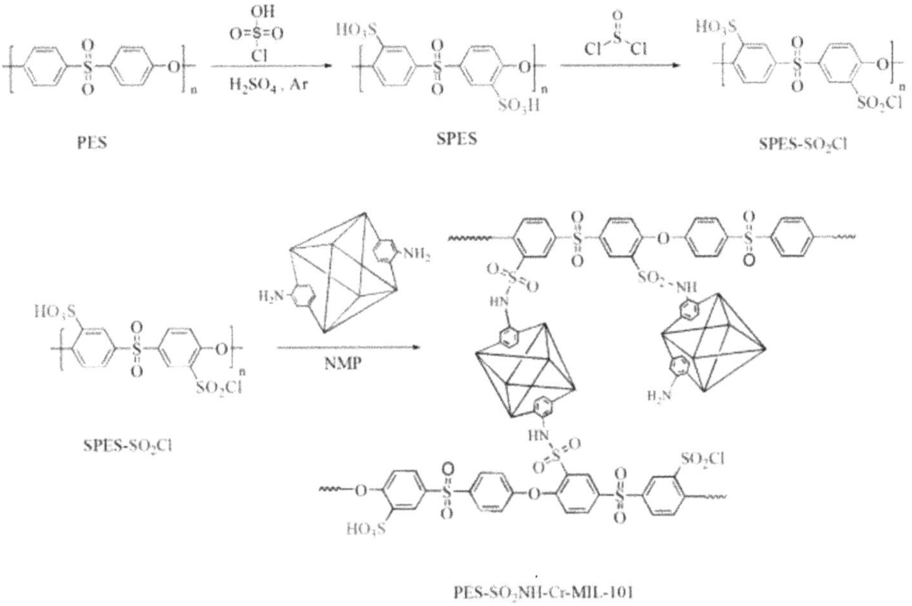

Figure 3. Schematic of the procedure for the synthesis of PES-SO_2Cl with Cr-MIL-101_2metal organic framework (MOF), utilized as PA-doped PEM for HTPEMFC [59].

Table 5 provides asummary ofseveral improvements to the sulfone group-containing PES, PSF, and phenyl sulfone (PSU). There are similarities inthe performance of sulfone-containing PEMs when compared to that of ketone-containing PEMs, where DS and IEC affect the water uptake and conductivity. It is also worth noting that the ketone and sulfone groups can be combined into one polymeric chain, which has beenreported in several studies. An example is the recent study conductedby J. Xu et al. [61] on crosslinked sulfonated poly(aryl ether ketone sulfone) (C-SPAEKS) with multiple sulfonic acid side chains. The existence of interaction by crosslinking restricted swelling and controlled the methanol permeability, whereas the IEC being larger than Nafion 117 offered better conductivities. Despite the positive outcome forkey properties, the power density was lower than that of Nafion 117, which may be due to the compatibility of the membrane within the components in the MEA.

Table 5. Uptake and electrochemical properties of PES- and PSF-type PEMs.

					PES- and PSF-type				
Membrane	**Year**	**%DS**	**Modifications**	**Fuel Cell Type**	**Filler Content**	**IEC (meq/g)**	**Water Uptake (%)**	**Proton Conductivity (S/cm)**	**Peak Power Density (mW/cm²)**
SPES-PBI [53]	2015	-	Ionic crosslinked with p-PBI	PEMFC	3wt% p-PBI	1.46	42.9 (80°C)	0.21 (80°C, 100%RH)	-
Imidazolium PSF[62]	2015	-	PSF with imidazolium pendants	HTPEMFC	-	-	-	0.04 (180°C, 0%RH)	269 (160°C, 0%RH, H_2/O_2)
SPES/CNW [54]	2016	-	Chitin nanowhiskers	PEMFC	7 wt% CNW	-	~19 (80°C)	~0.014 (80°C,100%RH)	-
SPES/NPHC [63]	2016	35	N-phythaloyl chitosan blend	DMFC	1wt% NPHC	1.29	41.5 (80°C)	0.0121 (80°C)	-
SPSF-SGO [64]	2017	71.55	Sulfonated graphene oxide	DMFC	3 wt% SGO	-	22.33	0.00427 (RT, 100%RH)	-
dsPFES-imPES[65]	2017	100	Sulfonated poly(fluorenyl ether sulfone)/imidazolium PES blend	PEMFC	2wt% imPES	1.17	89.7 (80°C)	0.35 (80°C,100%RH)	-
SPPSU/EG [55]	2018	2.24	SPPSU crosslinked with ethylene glycol (EG)	PEMFC	12 molecule EG/rpu	2.79	199 (RT)	0.23 (120°C,90%RH)	-
SPES-MOF [59]	2018	19	PES-SO_2Cl/Cr-MIL-101-NH_2 MOF	HTPEMFC	0.1 g MOF	3.18	35 (80°C)	0.041 (160°C, 0%RH)	238 (160°C, 0%RH, H_2/O_2)
PES-PVP [60]	2018	-	PES-PVP/graphitic carbon nitride (CN) nanosheets	HTPEMFC	0.5 wt% CN	-	-	0.12 (180°C, 0%RH)	634 (180°C, 0%RH, H_2/O_2)
SPES/S-MCM-41 [57]	2018	-	S-MCM-41 silica	PEMFC	2 wt% S-MCM-41	1.4	21.76	0.0694	-
SPSF/CW-Ser [58]	2019	40	Serine-modified cellulose nanowhiskers	DMFC	10wt% CW-Ser	-	~65 (80°C)	0.234 (80°C)	73.757 (60°C, 100%RH, 2M MeOH/O_2)
SPSU/SPPSU [56]	2019	-	Multiblock copolymer SPSU/SPPSU	PEMFC	1:9 (PSU:TMSCS ratio)	1.58	31.2 (60°C)	0.025 (80°C, 95%RH)	400 (70°C,100%RH, H_2/O_2)
Am-SPAEKS/C-SPAEKS [61]	2019	-	Crosslinked SPAEKS with multiple sulfonic acid groups	PEMFC	2 molar ratio of AMPS to Am-SPAEKS-DBS	2.09	14.6 (80°C	0.135 (80°C, 100%RH)	121.09 (80°C,100%RH, H_2/air)

2.4. *Polybenzimidazole (PBI)*

Mechanically and thermally strong polybenzimidazole (PBI) is of interest foruse as a PEM in HTPEMFC that operates in the temperature range of120 to 180 °C. Unlike PAEK-, PI-, or PES-type polymers that require additional functionalization to allow binding to PA molecules, the benzimidazole rings that naturally exist in the PBI backbone play this important role, enabling proton conduction to take place via both the Grotthuss and vehicular mechanism, between PA molecules (free and bounded to the –NH of benzimidazoles) and water molecules. The key properties, including the acid doping level (ADL) (or percentage PA uptake (%PA)), mechanical and thermal strengths, and proton conductivity, change with respective PBI types. The proton conductivity of PA-doped PBI membranes depends on the ADL, where more acid retained in the membrane leads toincreased conductivity. At a high ADL, the membrane's tensile strength is lowered due to the plasticizing effect of PA, despite the better conductivity achieved. Furthermore, as PA remains in liquid-form embedded within the membrane's free volume, leaching occurs over its lifetime of usage. In the MEA, PA leakage appearsto be more prominent on the cathode side of the membrane; in which water vapor produced from the cathode reaction canfacilitate the removal of PA. The leakage rate increases at a higher current density. Acid loss subsequently leads tohigher cell resistance and conductivity drop [66]. It is worth mentioning that several types of PBI, namely meta-PBI and ABPBI, dissolve poorly in common organic solvents, such as dimethylacetamide (DMAc), dimethylformamide (DMF), and N-methyl-2-pyrollidone (NMP), affecting the processability and membrane formation, due to their rigid structure and high glass transition temperature [67]. However, alternative acidic solvents may be utilized. N. Ratikanta et al. [68] studied the effect of methane sulfonic acid (MSA), trifluoroacetic acid (TFA), formic acid (FA), and sulfuric acid (SA) as solvents for ABPBI. The polymer casted from TFA displayed the highest PA absorption and proton conductivity at 150 °C, but had the weakest mechanical strength. The MSA-casted membrane offered the best cell performance with a small electrolyte resistance.

Improvements to PBI-Type PEMs

Under HTPEMFC conditions, the PEM is subjected to a faster rate of thermal and chemical degradation, mechanical stress, and PA loss. Researchers have followed similar strategies to improve the properties of PA-doped PBI similar to that of other hydrocarbon-based ion exchange membranes, aiming to balance the acid uptake and proton conductivity with their physical properties, as well asovercome the solubility issue and slow down the acid leaching rate. Aside from meta (m)- or para (p)-PBI and short-chained ABPBI, several PBI-based polymers with varying backbone structures have been synthesized in past studies. By the polymerization of different monomers, sulfonated PBI (SPBI), pyridine PBI (Py-PBI), PBI with ether bonds (OPBI), fluorinated PBI (F6-PBI), and branched PBI each have specific properties. Some have exhibited a better solubility, PA uptake, mechanical properties, and proton conductivity compared to m-PBIs [67,69]. They may also be modified further by crosslinking, introducing additional side-chains, blending, filler additives, and so on. Recently, the effects of new materials and modification techniques have been investigated.

X. Li et al. [70] grafted additional benzimidazole groups onto the backbone of aryl-ether PBI (Ph-PBI) through an N-substituted reaction without catalysts. The polymer solubilities of Ph-PBI and grafted Ph-PBI wereexcellent for most of the common solvents. The ADL and proton conductivity increased with a highergrafting degree. At 200 °C, the conductivity of the Ph-PBI with the maximum grafting degree reached 0.235 S/cm. However, the tensile stress wasparticularly low for the membrane due to the high ADL (3.2 MPa, slightly higher than OPBI, which was 2.5 MPa). For operation under the temperature concerned, it is important to consider how stable the membrane would be in the long term. H. Chen et al. [71] proposed a concept of dual proton transfer from a crosslinked membrane consisting of PBI with proton-donating and -accepting properties with polymeric ionic liquid (PIL) based on BuI-PBI and anion BF^- as proton acceptors. The anionic part of PIL wasable to accept protons to enhance the PA capacity, at the same time accepting protons through electrostatic interaction. The anions of PIL facilitated proton transfer; therefore, the conductivity increased. In contrast, the PIL caused a slight reduction in Young's modulus and the ultimate tensile strength (undoped), likely due to the destruction of hydrogen bonds in PBI during the N-quaternization reaction, forming PIL. With the inclusion of crosslinking, some –NH sites were important for PA absorption and proton transfer. L. Wang et al. [72] attempted tospare the –NH sites from the crosslinking reaction by the synthesis of branched F6-PBI with bis(3-phenyl-3,4-dihydro-2H-1,3-benzoxazinyl) isopropene (BA-a) as the polymeric covalent crosslinker. The unrestricted –NH sites, branched structure, and amine groups of crosslinkers helped in PA absorption and retention. The stability wasmaintained as a more rigid membrane was produced as a result of crosslinking; beneficial for the membrane in terms of itretaining its strength in a doped state and at high temperatures. Moreover, the membrane achieved a 690 mW/cm^2 peak power density at 160°C in H_2/O_2 HTPEMFC MEA tests. Long-term stability tests for 200 h also recorded stable open circuit voltage (OCV)(Figure 4), owing to the low H_2 permeability and internal resistance, high stability, and acceptable oxidative stability.

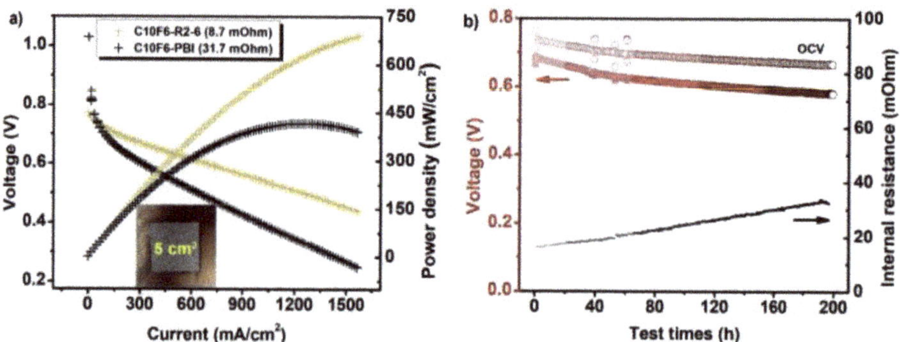

Figure 4. Outcomes of the crosslinked fluorinated PBI (F6-PBI) with unrestricted –NH sites (**a**) and the change in OCV of the membrane during a 200 h durability test, under 160°C, and in an anhydrous state (**b**) [73].

Conducting multiblock copolymerization to produce strong, phase-separated membranes is common among aryl ether ketone- and sulfone-type polymers. Recent attempts have been made to produce PBI-type block copolymers. L. Wang et al. [74] synthesized block copolymers consisting of varying ratios of OPBI and p-PBI segments. The resulting copolymer showed a nanophase-separated morphology that provided a larger free volume; hence, at equal ratios of OPBI and p-PBI, the ADL value was7.9 PA/rpu, which was higher than that of the individual segments (5.8 PA/rpu for OPBI and 4.7 PA/rpu for p-PBI). Furthermore, the equal ratios in the block copolymers exhibited the maximum phase separation degree. Therefore, there was alarge continuous channel for proton transfer. In turn, the proton conductivity achieved is stated to be five times higher than that of the individual segment. The peak power density reached 360 mW/cm^2, which is an obvious improvement (p-PBI: 250 mW/cm^2 and OPBI: 268 mW/cm^2) under H_2/air at 160°C, in an anhydrous state. Although there are advantages on the electrochemical side, the phase separation in a copolymer's microstructure may cause the membrane to be more susceptible to cracks because of the presence of nanocracks. This will make the membrane more likely to rupture under continuous stress.

Compatible fillers like silica, metal oxides, and graphene oxide are still utilized to enhance PBI properties. Functionalized fillers potentially provide stronger intermolecular interactions to minimize swelling due to a high PA uptake, while maintaining the conductivity. E. Abouzari-Lotf et al. [73] have shownthe enhancement of PA retention and proton conductivity of 2,6-Py-PBI phosphonic acid-functionalized graphene oxide (PGO). The PGO appeared to reduce the crystallinity of the polymer matrix while being able to maintain a sufficient mechanical strength. The more amorphous structure, which, according to the authors, provided stronger sites for PA retention and proton hopping pathways, increased the conductivity of the composite. The conductivity was also observed to increase with a small level of humidity (10% RH), where it was thought that the humidity reduced the contact resistance between the gas diffusion layer (GDL) and the membrane. A slower drop in conductivity was observed for the composites with PGO, shown in a 20 h durability test at 140 °C, without hydration, indicating a potentially stable membrane.

Table 6 provides asummary ofPBI-type membranes. Even recent studies have highlighted the advantages of PA-doped PBI membranes as solid electrolytes for HTPEMFCs that operate above the boiling point of water. At an appropriate ADL, it is feasible for PBIs to reach conductivities similar to those of Nafion in a temperature range of 160–200°C. Similarly, for the case of water-retaining SPEEK, SPI, and SPES, strengthening the PBI molecular chain interactions to minimize swelling and the risk of fast disintegration at a large ADL and high temperatures is necessary, especially when a high proton conductivity is desired. On the other hand, effective single-cell performance of the membranes would also rely on the PBI's compatibility with the MEA components, gas permeability, oxidative stability, and degradation rate over its lifetime.

Table 6. Uptake, acid doping level (ADL), and electrochemical properties of PBI-type PEMs.

PBI-type								
Membrane	**Year**	**Modifications**	**Fuel Cell Type**	**Filler Content**	**PA Uptake (%)**	**ADL (PA/rpu)**	**Proton Conductivity (S/cm)**	**Peak Power Density (mW/cm^2)**
PBI-4BPO$_x$[75]	2015	Boron phosphate	HTPEMFC	4 mole BPO_x/mole PBI	-	-	0.045 (150°C,5%RH)	~500 (150°C,0%RH,H_2/air)
Ph-PBI [76]	2016	Phenyl pendants	HTPEMFC	-	-	19.1 (160°C, 108 h)	0.138 (200°C,0%RH)	279 (160°C,0%RH,H_2/air)
Me-PBI [76]		Methylphenyl pendants				17.6 (160°C, 108 h)	0.123 (200°C,0%RH)	320 (160°C,0%RH,H_2/air)
PBIOH-ILS [77]	2017	Ionic liquid-functionalized silica	HTPEMFC	5% ILS	-	9.65 (110°C, 72 h)	0.106 (170°C,0%RH)	-
PBI-GO [78]	2017	Graphene oxide	HTPEMFC	2wt% GO	-	12 (336 h)	0.1704 (180°C, 0%RH)	380 (165°C, 0%RH, H_2/Air)
P-b-O-PBI [73]	2018	p-PBI/OPBI multiblock copolymer	HTPEMFC	0.5:0.5 (p-PBI:OPBI)	-	7.9 (80°C)	0.1 (180°C, 0%RH)	360 (160°C,0%RH, H_2/Air)
s-PBI [79]	2018	Azide naphthalene sulfonic acid-PBI	PEMFC	40wt% azide	-	-	0.006593 (RT, 0%RH)	-
PBI/lignin [80]	2018	Lignin	HTPEMFC	20wt% lignin	-	27 (RT, 24 h)	0.152 (160°C, 0%RH)	-
PBI-RGO/PPBI/PPBI-RGO [81]	2018	Radiation grafted sulfonated GO-PBI/Porous PB I three layer membrane	HTPEMFC	80% PPBI	500	20.4 (80°C, 48 h)	0.1138 (170°C, 0%RH)	-
g-PBI [70]	2018	Ph-PBI grafted with benzimidazolyl pendants	HTPEMFC	20% grafting degree	-	22.1 (120°C, 72 h)	0.212 (200°C, 0%RH)	443 (160°C, 0%RH, H_2/O_2)
ABPBI/S-Sep [82]	2019	Sulfonated sepiolite	HTPEMFC	2 wt% S-Sep	-	~3.5 (RT, 72 h)	0.051 (180°C, 0%RH)	230 (180°C, 0%RH, H_2/O_2)
cPBI-BF$_4$ [71]	2019	Crosslinked PBI with PBI-BuI PIL	HTPEMFC	40wt% PIL	362.5	19.7	0.117 (170°C, 0%RH)	-
CF6PBI-R2-6 [72]	2019	Crosslinked branched F6-PBI with BA-a	HTPEMFC	-	~69.5 (120°C)	-	~0.07 (180°C, 0%RH)	690 (160°C, 0% RH, H_2/O_2)
2,6-Py-PBI/PGO [73]	2019	Phosphonated graphene oxide	HTPEMFC	1.5 wt% PGO	-	5.8 (45°C, 168 h)	0.0764 (140°C, 0%RH)	359 (120°C, 0%RH, H_2/Air)

2.5. Polyphenylene Oxide (PPO)

Polyphenylene oxides (PPOs), also referred to as polyphenylene ethers (PPEs), are ether-containing aromatic polymerswithelectrical insulation properties, high mechanical strengths, and good chemical resistance. In terms of their thermal aspects, the glass transition temperatures of PPO may reach as high as 210°C, although this varies, depending on their grade and modifications. The excellent dimensional stability of PPO is related to itslow moisture absorption, even when exposed to boiling water [83]. Modified PPO or PPE-based AEMs for fuel cells have beenwell-studied in past years; however, several recent studies have also reported the potential use of modified PPOs as PEMs.

Improvements to PPO-Type PEMs

Similar to other sulfonated aromatic PEMs, SPPO's IEC, water uptake, and conductivity relieson its degree of sulfonation. However, the weaker thermal stability and mechanical strength of the SPPO membrane may shorten its lifetime. I. Petreanu et al. [84] incorporated silica particles into the SPPO matrix and reported a higher ultimate tensile strength for the silica-containing membrane in a hydrated state, compared to pristine SPPO. Other key properties, including IEC and water uptake, lowered in the presence of silica. Beside SPPOs, the bromomethylated PPOs (BPPOs) with hydrophobic properties caneffectively improve the methanol resistance and DMFC cell performance of SPEEK, as investigated by X. Liu et al. [85], who employed a SPEEK-BPPO blend membrane. While the water uptake and conductivity wereaffected in the presence of BPPO, the excellent methanol resistance and selectivity contributed to the better DMFC single-cell performance at a 5M methanol concentration, achieving a power density five times higher than that of Nafion 117.

Recently, there havealso been attempts toutilize PPOs atoperating temperatures above 100 °C. X. Zhu et al. [86] prepared a crosslinked methylimidazole-functionalized PPO incorporated with phosphonic acid-functionalized siloxane as a PEM for high-temperature and low-humidity conditions. The effective proton conductive properties of the functionalized siloxane allowed the conductivity to further elevate after 100 °C at 5% RH, while crosslinking strengthened the membrane's mechanical properties and oxidative stability. This suggests that a crosslinker functionalized with effective proton conductive functional groups has a positive role in simultaneously enforcing the membrane and improving the electrochemical properties. Furthermore, an investigation of a crosslinked PPO containing a triazole side chain, synthesized by J. Jang et al. [87], showed that the highly crosslinked structure wasable to minimize PA leaching, whilstimproving the mechanical and thermal strength, but also suppressed high PA absorption, leading to lower conductivities. The addition of more triazole side chains raises the PA uptake. Since the PA uptake still has a significant effect on theproton conductivity, the membrane must retain a suitable amount of PA for better conductivities. Therefore, this would require the optimization of both the degree of crosslinking and triazole content to balance out the key aspects to function as effective PEM.

Table 7 summarizes several of the recently investigated PPO-type PEMs. While the number of studies for PPO-type fuel cell PEMs seems limited compared to AEMs, there is still good potential for the utilization of PPO as membrane material. Much like the other aromatic-based membranes, the optimization of individual aspects of the membrane is an important step in achieving electrochemical and durability balance.

Table 7. Water uptake, PA uptake, and the electrochemical properties of PPO-type PEMs.

PPO-type									
Membrane	**Year**	**Modifications**	**Fuel Cell Type**	**Filler Content**	**IEC(meq/g)**	**Water Uptake (%)**	**PA Uptake (%)**	**Proton Conductivity (S/cm)**	**Peak Power Density (mW/cm²)**
SPPO-HGM-SPPO [88]	2015	Hollow glass microspheres (HGMs)	DMFC	9wt% HGM	2.164	19.31	-	0.0318 (20°C, 100%RH)	81.5 (RT, 2M MeOH/O_2)
PPO-MeIM[89]	2017	Methylimidazolium PPO	HTPEMFC	4:10 (MeIM:BPPO)	-	-	135 (30°C, 24 h)	0.0679 (160°C, 0% RH)	280 (160 °C, 0%RH, H_2/O_2)
SPEEK/BPPO [85]	2017	SPEEK/BPPO blend	DMFC	20wt% BPPO	1.21 (mmol/g)	11.76	-	0.064 (60°C, 100%RH)	23.9 (60 °C, 10M MeOH/O_2)
SPPO+TEOS [84]	2017	TEOS-based silica nanoparticles	PEMFC	-	1.75	66	-	-	-
QPPO-MIm/ ATMP-APTES [90]	2019	Phosphonic acid-functionalized siloxane	HTPEMFC	15wt% ATMP-APTES	1.04 (mmol/g)	38.91 (80°C)	-	0.0848 (160°C, 5%RH)	638 (160 °C, 5%, H_2/O_2)
XTPPO [87]	2020	Crosslinked triazole PPO	HTPEMFC	40% bromination degree, 10% degree of crosslinking	-	-	211 (120°C, 15 h)	0.064 (180 °C, 0%RH)	-

3. Challenges and Future Perspectives

As seen from the extensive research conductedin past and recent years, PEMs derived from various hydrocarbon-based polymers hold a lot of potentials to be applied as solid electrolyte alternatives to Nafion for LTPEMFCs, HTPEMFCs, and DMFCs. The different modification methods employed to enhance these PEMs have observed that improvements intheir electrochemical characteristics and durability are possible, taking into account thecontrol, adjustment, and optimization of individual key properties. While several studies have recorded the performance of hydrocarbon-based membranes and their modified forms in long-term MEA tests, knowledge on characteristic changes in long-term durability under fluctuating temperatures, pressures, and fuel flows in actual fuel cell stacks could be more comprehensive.

Nafion has been widely commercialized as it has a role as the standard in fuel cell systems and their production has been achieved at anindustrial scale. PEEK, PES, PI, PBI, and PPO polymers are already manufactured atlarge scales to cater for their diverse applications besides PEMs. However, suitable proton or ion conducting PEEK, PES, PI, PBI, and PPO specifically for fuel cell applications have yet to enter the competitive market alongside Nafion. The difficulty incommercializing these alternative PEMs is due to the durability issue and stability related to their electrochemical properties, which still requires further optimization. Hydrocarbon-based PEMs synthesized from individual monomers may hold some advantages in long-term durability compared to those derived from commercial polymers. However, the polymerization reaction process can be complex, andthe stoichiometric ratios, reaction conditions, purification, recovery, and film formation process must be specified at an upscale level, sothe economic feasibility needs consideration. Using commercial polymers that undergo functionalization reactions, blending, or filler inclusion may be a more viable option. Again, the functionalization level (ex: DS), MEA compatibility, and durability should be optimized. The cost of secondary materials (blended polymer, fillers, etc.) also needs to be included.

4. Conclusions

In summary, aromatic-based membranes consisting of strong aryl rings, ether, ketones, sulfones, imides, and benzimidazole linkages, along with the optional inclusion of fluorinated structures, followed by functionalization with strong ionic conducting groups, provide the membrane with the mechanical, thermal, and chemical strengths; water or PA retention abilities; and ionic/protonic conductive properties required for them to function as PEMs for fuel cells operating in the temperature range fromambient to around 200°C. Furthermore, the low methanol and gas permeation of these alternative PEMs even offer benefits toward better MEA performances. Larger IEC and water absorption of high DS SPEEK, SPES, and SPI would lead to higher proton conductivities, that would otherwise be lower than those of Nafion due to smaller hydrophobic/hydrophilic phase separation and the less effective ion transport channel of hydrocarbon-based PEMs. On the other hand, high PA uptake or ADL- of PBI-type membranes, as well as imidazole or ionic liquid-functionalized PAEK, PES, Pi, or PPO, facilitate conductivity under anhydrous conditions. However, the excessive accumulation of water or PA molecules within the free volumes of these polymers can cause large swelling, deterioration in the mechanical strength, easier fuel permeation, and faster degradation that may becaused bypoor oxidative stability, which is adisadvantage in terms ofthe hydrocarbon-based PEM's durability, thus affecting their lifetime under fuel cell operating conditions.

To this day, there have been various strategies and methods adopted by researchers to enhance the PEM characteristics of aromatic-based polymers. Crosslinking, multiblock-copolymerization, the introduction of inorganic/organic fillers/nanofillers, the addition of branching or pendant structures, and morphological modifications through the inclusion of nanofibers within the polymer matrix have beenproven to improve the mechanical, thermal, oxidative stability, fuel resistance, and electrochemical performance of these alternative PEMs, especially concerning the utilization of the benefits of PEMs with a high DS and water/PA uptake. However, some challenges still remain for these PEMs, even as they have been further modified. Crosslinking could lead to brittleness. Filler/nanofiller addition may form more tortuous proton conductive pathways. The compatibility of these modified aromatic-based PEMs forthe MEA components may be different fromthat of Nafion and more elaborate investigations are needed, such as explorations on the electrolyte-electrode contact, catalyst quantity, and flowrates of reactants and oxidants. Nevertheless, aromatic-based PEMs still hold great potential as effective and low-cost alternative PEMs for fuel cells. The success of producing PEMs with excellent performances dependson the balance between the electrochemical properties, physical characteristics, MEA compatibility, and durability, which requires a careful in-depth understanding of the fundamental characteristics of the polymers and the optimization of individual aspects of the membrane.

Author Contributions: Conceptualization, R.W. and W.W.Y.; methodology, W.W.Y and R.W.; investigation, R.R.R.S.; resources, K.M. and P.J.; writing—original draft preparation, R.R.R.S.; writing—review and editing, W.W.Y, R.W. and K.M.; supervision, W.W.Y., R.W., K.M. and P.J.; project administration, W.W.Y., R.W. and K.M., funding acquisition, R.W. and W.W.Y. All authors have read and agreed to the published version of the manuscript.

Acknowledgments: The authors would like to acknowledge Taylor's University's Masters Scholarship Programme. Furthermore, we want to extend our acknowledgement to UniversitiKebangsaan Malaysia through the project code DIP-2018-012 for supporting this research.

Conflicts of Interest: The authors declare no conflict of interest.The authors also declare that the funders had no role in the design of the study; in the collection, analyses, or interpretation of data; in the writing of the manuscript, or in the decision to publish the results.

References

1. Mekhilef, S.; Saidur, R.; Safari, A. Comparative study of different fuel cell technologies. *Renew. Sustain. Energy Rev.* **2012**, *16*, 981–989. [CrossRef]
2. Zhang, L.; Chae, S.-R.; Hendren, Z.; Park, J.-S.; Wiesner, M.R. Recent advances in proton exchange membranes for fuel cell applications. *Chem. Eng. J.* **2012**, *204–206*, 87–97. [CrossRef]
3. Wang, Y.; Chen, K.S.; Mishler, J.; Cho, S.C.; Adroher, X.C. A review of polymer electrolyte membrane fuel cells: Technology, applications, and needs on fundamental research. *Appl. Energy* **2011**, *88*, 981–1007. [CrossRef]
4. Peighambardoust, S.J.; Rowshanzamir, S.; Amjadi, M. Review of the proton exchange membranes for fuel cell applications. *Int. J. Hydrogen Energy* **2010**, *35*, 9349–9384. [CrossRef]
5. Kim, D.J.; Jo, M.J.; Nam, S.Y. A review of polymer–nanocomposite electrolyte membranes for fuel cell application. *J. Ind. Eng. Chem.* **2015**, *21*, 36–52. [CrossRef]
6. Authayanun, S.; Im-orb, K.; Arpornwichanop, A. A review of the development of high temperature proton exchange membrane fuel cells. *Chin. J. Catal.* **2015**, *36*, 473–483. [CrossRef]
7. Bakangura, E.; Wu, L.; Ge, L.; Yang, Z.; Xu, T. Mixed matrix proton exchange membranes for fuel cells: State of the art and perspectives. *Prog. Polym. Sci.* **2016**, *57*, 103–152. [CrossRef]
8. Kang, H.; Hong, J.; Hong, T.; Han, D.; Chin, S.; Lee, M. Determining the optimal long-term service agreement period and cost considering the uncertain factors in the fuel cell: From the perspectives of the sellers and generators. *Appl. Energy* **2019**, *237*, 378–389. [CrossRef]
9. Leong, J.X.; Daud, W.R.W.; Ghasemi, M.; Liew KBen Ismail, M. Ion exchange membranes as separators in microbial fuel cells for bioenergy conversion: A comprehensive review. *Renew. Sustain. Energy Rev.* **2013**, *28*, 575–587. [CrossRef]
10. Branco, C.M.; Sharma, S.; de Camargo Forte, M.M.; Steinberger-Wilckens, R. New approaches towards novel composite and multilayer membranes for intermediate temperature-polymer electrolyte fuel cells and direct methanol fuel cells. *J. Power Sources* **2016**, *316*, 139–159. [CrossRef]
11. Marcinkoski, J.; Spendelow, J.; Wilson, A.; Papageorgopoulos, D.; Ahluwalia, R.; James, B.; Houchins, C.; Huya-Kouadio, J. *DOE Hydrogen and Fuel Cells Program Record*; DOE: Washington, DC, USA, 2017. Available online: https://www.hydrogen.energy.gov/pdfs/15006_separation_distance_reduction.pdf (accessed on 26 March 2020).
12. Kausar, A. Progression from Polyimide to Polyimide Composite in Proton-Exchange Membrane Fuel Cell: A Review. *Polym. Plast. Technol. Eng.* **2017**, *56*, 1375–1390. [CrossRef]
13. Knauth, P.; Di Vona, M.L. Sulfonated aromatic ionomers: Analysis of proton conductivity and proton mobility. *Solid State Ion.* **2012**, *225*, 255–259. [CrossRef]

14. Park, J.-S.; Shin, M.-S.; Kim, C.-S. Proton exchange membranes for fuel cell operation at low relative humidity and intermediate temperature: An updated review. *Curr. Opin. Electrochem.* **2017**, *5*, 43–55. [CrossRef]
15. Ogungbemi, E.; Ijaodola, O.; Khatib, F.N.; Wilberforce, T.; El Hassan, Z.; Thompson, J.; Ramadan, M.; Olabi, A.G. Fuel cell membranes—Pros and cons. *Energy* **2019**, *172*, 155–172. [CrossRef]
16. Yan, X.; He, G.; Wu, X.; Benziger, J. Ion and water transport in functionalized PEEK membranes. *J. Membr. Sci.* **2013**, *429*, 13–22. [CrossRef]
17. Parnian, M.J.; Rowshanzamir, S.; Gashoul, F. Comprehensive investigation of physicochemical and electrochemical properties of sulfonated poly (ether ether ketone) membranes with different degrees of sulfonation for proton exchange membrane fuel cell applications. *Energy* **2017**, *125*, 614–628. [CrossRef]
18. Iulianelli, A.; Basile, A. Sulfonated PEEK-based polymers in PEMFC and DMFC applications: A review. *Int. J. Hydrogen Energy* **2012**, *37*, 15241–15255. [CrossRef]
19. Portale, G.; Carbone, A.; Martinelli, A.; Passalacqua, E. Microstructure, state of water and proton conductivity of sulfonated poly(ether ether ketone). *Solid State Ion.* **2013**, *252*, 62–67. [CrossRef]
20. Park, C.H.; Lee, C.H.; Sohn, J.-Y.; Park, H.B.; Guiver, M.D.; Lee, Y.M. Phase Separation and Water Channel Formation in Sulfonated Block Copolyimide. *J. Phys. Chem. B* **2010**, *114*, 12036–12045. [CrossRef]
21. Park, C.H.; Kim, T.-H.; Nam, S.Y.; Hong, Y.T. Water channel morphology of non-perfluorinated hydrocarbon proton exchange membrane under a low humidifying condition. *Int. J. Hydrogen Energy* **2018**. [CrossRef]
22. Shaari, N.; Kamarudin, S.K. Graphene in electrocatalyst and proton conductiong membrane in fuel cell applications: An overview. *Renew. Sustain. Energy Rev.* **2017**, *69*, 862–870. [CrossRef]
23. Farooqui, U.R.; Ahmad, A.L.; Hamid, N.A. Graphene oxide: A promising membrane material for fuel cells. *Renew. Sustain. Energy Rev.* **2018**, *82*, 714–733. [CrossRef]
24. Ying, Y.P.; Kamarudin, S.K.; Masdar, M.S. Silica-related membranes in fuel cell applications: An overview. *Int. J. Hydrogen Energy* **2018**, *43*, 16068–16084. [CrossRef]
25. Kreuer, K.D. On the development of proton conducting polymer membranes for hydrogen and methanol fuel cells. *J.Membr. Sci.* **2001**, *185*, 29–39. [CrossRef]
26. Banerjee, S.; Kar, K.K. Impact of degree of sulfonation on microstructure, thermal, thermomechanical and physicochemical properties of sulfonated poly ether ether ketone. *Polymer (Guildf.)* **2017**, *109*, 176–186. [CrossRef]
27. Mirfarsi, S.H.; Karimi, A.; Rowshanzamir, S.; Parnian, M.J. Study of mechanical degradation of sulfonated poly (ether ether ketone) membrane using ex-situ hygrothermal cycles for polymer electrolyte fuel cell application. *J. Power Sources* **2018**, *401*, 73–84. [CrossRef]
28. Karimi, A.; Kalfati, M.S.; Rowshanzamir, S. Investigation, modeling, and optimization of parameters affecting sulfonated polyether ether ketone membrane-electrode assembly. *Int. J. Hydrogen Energy* **2019**, *44*, 1096–1109. [CrossRef]
29. Liu, X.; He, S.; Liu, S.; Jia, H.; Chen, L.; Zhang, B.; Zhang, L.; Lin, J. The roles of solvent type and amount of residual solvent on determining the structure and performance of sulfonated poly(ether ether ketone) proton exchange membranes. *J. Membr. Sci.* **2017**, *523*, 163–172. [CrossRef]
30. Gao, S.; Xu, H.; Fang, Z.; Ouadah, A.; Chen, H.; Chen, X.; Shi, L.; Ma, B.; Jing, C.; Zhu, C. Highly sulfonated poly(ether ether ketone) grafted on graphene oxide as nanohybrid proton exchange membrane applied in fuel cells. *Electrochim. Acta* **2018**, *283*, 428–437. [CrossRef]
31. Bano, S.; Negi, Y.S.; Illathvalappil, R.; Kurungot, S.; Ramya, K. Studies on nano composites of SPEEK/ethylene glycol/cellulose nanocrystals as promising proton exchange membranes. *Electrochim. Acta* **2019**, *293*, 260–272. [CrossRef]
32. Liu, G.; Tsen, W.C.; Jang, S.C.; Hu, F.; Zhong, F.; Liu, H.; Wang, G.; Wen, S.; Zheng, G.; Gong, C. Mechanically robust and highly methanol-resistant sulfonated poly(ether ether ketone)/poly(vinylidene fluoride) nanofiber composite membranes for direct methanol fuel cells. *J.Membr. Sci.* **2019**, *591*, 117321. [CrossRef]
33. Wu, Y.; He, G.; Wu, X.; Yuan, Q.; Gong, X.; Zhen, D.; Sun, B. Confinement of functionalized graphene oxide in sulfonated poly (ether ether ketone) nanofibers by coaxial electrospinning for polymer electrolyte membranes. *Int. J. Hydrogen Energy* **2019**, *44*, 7494–7504. [CrossRef]
34. Kang, K.; Kim, D. Toughened polymer electrolyte membranes composed of sulfonated poly(arylene ether ketone) block copolymer and organosiloxane network for fuel cell. *Solid State Ion.* **2019**, *335*, 23–31. [CrossRef]

35. Li, J.; Wang, S.; Liu, F.; Tian, X.; Wang, X.; Chen, H.; Mao, T.; Wang, Z. HT-PEMs based on nitrogen-heterocycle decorated poly (arylene ether ketone) with enhanced proton conductivity and excellent stability. *Int. J. Hydrogen Energy* **2018**, *43*, 16248–16257. [CrossRef]
36. Yang, J.; Wang, Y.; Yang, G.; Zhan, S. New anhydrous proton exchange membranes based on fluoropolymers blend imidazolium poly (aromatic ether ketone)s for high temperature polymer electrolyte fuel cells. *Int. J. Hydrogen Energy* **2018**, *43*, 8464–8473. [CrossRef]
37. Wang, R.; Wu, X.; Yan, X.; He, G.; Hu, Z. Proton conductivity enhancement of SPEEK membrane through n-BuOH assisted self-organization. *J.Membr. Sci.* **2015**, *479*, 46–54. [CrossRef]
38. Kang, K.; Kim, D. Comparison of proton conducting polymer electrolyte membranes prepared from multi-block and random copolymers based on poly(arylene ether ketone). *J. Power Sources* **2015**, *281*, 146–157. [CrossRef]
39. Salarizadeh, P.; Javanbakht, M.; Pourmahdian, S.; Beydaghi, H. Influence of amine-functionalized iron titanate as filler for improving conductivity and electrochemical properties of SPEEK nanocomposite membranes. *Chem. Eng. J.* **2016**, *299*, 320–331. [CrossRef]
40. Mong, A.; Le Yang, S.; Kim, D. Pore-filling polymer electrolyte membrane based on poly (arylene ether ketone) for enhanced dimensional stability and reduced methanol permeability. *J. Membr. Sci.* **2017**, *543*, 133–142. [CrossRef]
41. Qiu, X.; Dong, T.; Ueda, M.; Zhang, X.; Wang, L. Sulfonated reduced graphene oxide as a conductive layer in sulfonated poly(ether etherketone) nanocomposite membranes. *J.Membr. Sci.* **2017**, *524*, 663–672. [CrossRef]
42. Che, Q.; Fan, H.; Duan, X.; Feng, F.; Mao, W.; Han, X. Layer by layer self-assembly fabrication of high temperature proton exchange membrane based on ionic liquids and polymers. *J. Mol. Liq.* **2018**, *269*, 666–674. [CrossRef]
43. Kowsari, E.; Zare, A.; Ansari, V. Phosphoric acid-doped ionic liquid-functionalized graphene oxide/sulfonated polyimide composites as proton exchange membrane. *Int. J. Hydrogen Energy* **2015**, *40*, 13964–13978. [CrossRef]
44. Zhang, B.; Ni, J.; Xiang, X.; Wang, L.; Chen, Y. Synthesis and properties of reprocessable sulfonated polyimides cross-linked via acid stimulation for use as proton exchange membranes. *J. Power Sources* **2017**, *337*, 110–117. [CrossRef]
45. Genies, C.; Mercier, R.; Sillion, B.; Petiaud, R.; Cornet, N.; Gebel, G.; Pineri, M. Stability study of sulfonated phthalic and naphthalenic polyimide structures in aqueous medium. *Polymer* **2001**, *42*, 5097–5105. [CrossRef]
46. Yao, Z.; Zhang, Z.; Hu, M.; Hou, J.; Wu, L.; Xu, T. Perylene-based sulfonated aliphatic polyimides for fuel cell applications: Performance enhancement by stacking of polymer chains. *J.Membr. Sci.* **2018**, *547*, 43–50. [CrossRef]
47. Liaqat, K.; Rehman, W.; Saeed, S.; Waseem, M.; Fazil, S.; Shakeel, M.; Kang, P. Synthesis and characterization of novel sulfonated polyimide with varying chemical structure for fuel cell applications. *Solid State Ion.* **2018**, *319*, 141–147. [CrossRef]
48. Feng, S.; Kondo, S.; Kaseyama, T.; Nakazawa, T.; Kikuchi, T.; Selyanchyn, R.; Fujikawa, S.; Christiani, L.; Sasaki, K.; Nishihara, M. Development of polymer-polymer type charge-transfer blend membranes for fuel cell application. *J. Membr. Sci.* **2018**, *548*, 223–231. [CrossRef]
49. Ito, G.; Tanaka, M.; Kawakami, H. Sulfonated polyimide nanofiber framework: Evaluation of intrinsic proton conductivity and application to composite membranes for fuel cells. *Solid State Ion.* **2018**, *317*, 244–255. [CrossRef]
50. You, P.Y.; Kamarudin, S.K.; Masdar, M.S. Improved performance of sulfonated polyimide composite membranes with rice husk ash as a bio-filler for application in direct methanol fuel cells. *Int. J. Hydrogen Energy* **2019**, *44*, 1857–1866. [CrossRef]
51. Yue, Z.; Cai, Y.-B.; Xu, S. Phosphoric acid-doped organic-inorganic cross-linked sulfonated poly(imide-benzimidazole) for high temperature proton exchange membrane fuel cells. *Int. J. Hydrogen Energy* **2016**, *41*, 10421–10429. [CrossRef]
52. Sastri, V.R. High-Temperature Engineering Thermoplastics: Polysulfones, Polyimides, Polysulfides, Polyketones, Liquid Crystalline Polymers, and Fluoropolymers. *Plast. Med. Devices* **2010**, 175–215. [CrossRef]
53. Won, M.; Kwon, S.; Kim, T.-H. High performance blend membranes based on sulfonated poly(arylene ether sulfone) and poly(p-benzimidazole) for PEMFC applications. *J. Ind. Eng. Chem.* **2015**, *29*, 104–111. [CrossRef]

54. Zhang, C.; Zhuang, X.; Li, X.; Wang, W.; Cheng, B.; Kang, W.; Cai, Z.; Li, M. Chitin nanowhisker-supported sulfonated poly(ether sulfone) proton exchange for fuel cell applications. *Carbohydr. Polym.* **2016**, *140*, 195–201. [CrossRef] [PubMed]
55. Matsushita, S.; Kim, J.-D. Organic solvent-free preparation of electrolyte membranes with high proton conductivity using aromatic hydrocarbon polymers and small cross-linker molecules. *Solid State Ion.* **2018**, *316*, 102–109. [CrossRef]
56. Ureña, N.; Pérez-Prior, M.T.; Del Río, C.; Várez, A.; Sanchez, J.Y.; Iojoiu, C.; Levenfeld, B. Multiblock copolymers of sulfonated PSU/PPSU Poly(ether sulfone)s as solid electrolytes for proton exchange membrane fuel cells. *Electrochim. Acta* **2019**, *302*, 428–440. [CrossRef]
57. Gahlot, S.; Sharma, P.P.; Yadav, V.; Jha, P.K.; Kulshrestha, V. Nanoporous composite proton exchange membranes: High conductivity and thermal stability. *Colloids Surf. A Physicochem. Eng. Asp.* **2018**, *542*, 8–14. [CrossRef]
58. Xu, X.; Zhao, G.; Wang, H.; Li, X.; Feng, X.; Cheng, B.; Shi, L.; Kang, W.; Zhuang, X.; Yin, Y. Bio-inspired amino-acid-functionalized cellulose whiskers incorporated into sulfonated polysulfone for proton exchange membrane. *J. Power Sources* **2019**, *409*, 123–131. [CrossRef]
59. Anahidzade, N.; Abdolmaleki, A.; Dinari, M.; Firouz Tadavani, K.; Zhiani, M. Metal-organic framework anchored sulfonated poly(ether sulfone) as a high temperature proton exchange membrane for fuel cells. *J.Membr. Sci.* **2018**, *565*, 281–292. [CrossRef]
60. Bai, H.; Wang, H.; Zhang, J.; Wu, C.; Zhang, J.; Xiang, Y.; Lu, S. Simultaneously enhancing ionic conduction and mechanical strength of poly(ether sulfones)-poly(vinyl pyrrolidone) membrane by introducing graphitic carbon nitride nanosheets for high temperature proton exchange membrane fuel cell application. *J. Membr. Sci.* **2018**, *558*, 26–33. [CrossRef]
61. Xu, J.; Zhang, Z.; Yang, K.; Zhang, H.; Wang, Z. Synthesis and properties of novel cross-linked composite sulfonated poly (aryl ether ketone sulfone) containing multiple sulfonic side chains for high-performance proton exchange membranes. *Renew. Energy* **2019**, *138*, 1104–1113. [CrossRef]
62. Yang, J.; Wang, J.; Liu, C.; Gao, L.; Xu, Y.; Che, Q.; He, R. Influences of the structure of imidazolium pendants on the properties of polysulfone-based high temperature proton conducting membranes. *J. Membr. Sci.* **2015**, *493*, 80–87. [CrossRef]
63. Muthumeenal, A.; Neelakandan, S.; Kanagaraj, P.; Nagendran, A. Synthesis and properties of novel proton exchange membranes based on sulfonated polyethersulfone and N-phthaloyl chitosan blends for DMFC applications. *Renew. Energy* **2016**, *86*, 922–929. [CrossRef]
64. Bunlengsuwan, P.; Paradee, N.; Sirivat, A. Influence of Sulfonated Graphene Oxide on Sulfonated Polysulfone Membrane for Direct Methanol Fuel Cell. *Polym. Plast. Technol. Eng.* **2017**, *56*, 1695–1703. [CrossRef]
65. Kwon, S.; Lee, B.; Kim, T.H. High performance blend membranes based on densely sulfonated poly(fluorenyl ether sulfone) block copolymer and imidazolium-functionalized poly(ether sulfone). *Int. J. Hydrogen Energy* **2017**, *42*, 20176–20186. [CrossRef]
66. Jeong, Y.H.; Oh, K.; Ahn, S.; Kim, N.Y.; Byeon, A.; Park, H.Y.; Lee, S.Y.; Park, H.S.; Yoo, S.J.; Jang, J.H.; et al. Investigation of electrolyte leaching in the performance degradation of phosphoric acid-doped polybenzimidazole membrane-based high temperature fuel cells. *J. Power Sources* **2017**, *363*, 365–374. [CrossRef]
67. Araya, S.S.; Zhou, F.; Liso, V.; Sahlin, S.L.; Vang, J.R.; Thomas, S.; Gao, X.; Jeppesen, C.; Kær, S.K. A comprehensive review of PBI-based high temperature PEM fuel cells. *Int. J. Hydrogen Energy* **2016**, *41*, 21310–21344. [CrossRef]
68. Nayak, R.; Sundarraman, M.; Ghosh, P.C.; Bhattacharyya, A.R. Doped poly (2, 5-benzimidazole) membranes for high temperature polymer electrolyte fuel cell: Influence of various solvents during membrane casting on the fuel cell performance. *Eur. Polym. J.* **2018**, *100*, 111–120. [CrossRef]
69. Haque, M.A.; Sulong, A.B.; Loh, K.S.; Majlan, E.H.; Husaini, T.; Rosli, R.E. Acid doped polybenzimidazoles based membrane electrode assembly for high temperature proton exchange membrane fuel cell: A review. *Int. J. Hydrogen Energy* **2017**, *42*, 9156–9179. [CrossRef]
70. Li, X.; Wang, P.; Liu, Z.; Peng, J.; Shi, C.; Hu, W.; Jiang, Z.; Liu, B. Arylether-type polybenzimidazoles bearing benzimidazolyl pendants for high-temperature proton exchange membrane fuel cells. *J. Power Sources* **2018**, *393*, 99–107. [CrossRef]

71. Chen, H.; Wang, S.; Li, J.; Liu, F.; Tian, X.; Wang, X.; Mao, T.; Xu, J.; Wang, Z. Novel cross-linked membranes based on polybenzimidazole and polymeric ionic liquid with improved proton conductivity for HT-PEMFC applications. *J. Taiwan Inst. Chem. Eng.* **2019**, *95*, 185–194. [CrossRef]
72. Wang, L.; Liu, Z.; Liu, Y.; Wang, L. Crosslinked polybenzimidazole containing branching structure with no sacrifice of effective N-H sites: Towards high-performance high-temperature proton exchange membranes for fuel cells. *J. Membr. Sci.* **2019**, *583*, 110–117. [CrossRef]
73. Abouzari-Lotf, E.; Zakeri, M.; Nasef, M.M.; Miyake, M.; Mozarmnia, P.; Bazilah, N.A.; Emelin, N.F.; Ahmad, A. Highly durable polybenzimidazole composite membranes with phosphonated graphene oxide for high temperature polymer electrolyte membrane fuel cells. *J. Power Sources* **2019**, *412*, 238–245. [CrossRef]
74. Wang, L.; Liu, Z.; Ni, J.; Xu, M.; Pan, C.; Wang, D.; Liu, D.; Wang, L. Preparation and investigation of block polybenzimidazole membranes with high battery performance and low phosphoric acid doping for use in high-temperature fuel cells. *J. Membr. Sci.* **2019**, *572*, 350–357. [CrossRef]
75. Mamlouk, M.; Scott, K. A boron phosphate-phosphoric acid composite membrane for medium temperature proton exchange membrane fuel cells. *J Power Sources* **2015**, *286*, 290–298. [CrossRef]
76. Li, X.; Ma, H.; Shen, Y.; Hu, W.; Jiang, Z.; Liu, B.; Guiver, M.D. Dimensionally-stable phosphoric acid–doped polybenzimidazoles for high-temperature proton exchange membrane fuel cells. *J. Power Sources* **2016**, *336*, 391–400. [CrossRef]
77. Liu, F.; Wang, S.; Li, J.; Tian, X.; Wang, X.; Chen, H.; Wang, Z. Polybenzimidazole/ionic-liquid-functional silica composite membranes with improved proton conductivity for high temperature proton exchange membrane fuel cells. *J. Membr. Sci.* **2017**, *541*, 492–499. [CrossRef]
78. Üregen, N.; Pehlivanoğlu, K.; Özdemir, Y.; Devrim, Y. Development of polybenzimidazole/graphene oxide composite membranes for high temperature PEM fuel cells. *Int. J. Hydrogen Energy* **2017**, *42*, 2636–2647. [CrossRef]
79. Huang, B.; Wang, X.; Fang, H.; Jiang, S.; Hou, H. Mechanically strong sulfonated polybenzimidazole PEMs with enhanced proton conductivity. *Mater. Lett.* **2019**, *234*, 354–356. [CrossRef]
80. Barati, S.; Abdollahi, M.; Khoshandam, B.; Mehdipourghazi, M. Highly proton conductive porous membranes based on polybenzimidazole/lignin blends for high temperatures proton exchange membranes: Preparation, characterization and morphology-proton conductivity relationship. *Int. J. Hydrogen Energy* **2018**, *43*, 19681–19690. [CrossRef]
81. Cai, Y.; Yue, Z.; Teng, X.; Xu, S. Radiation grafting graphene oxide reinforced polybenzimidazole membrane with a sandwich structure for high temperature proton exchange membrane fuel cells in anhydrous atmosphere. *Eur. Polym. J.* **2018**, *103*, 207–213. [CrossRef]
82. Zhang, X.; Liu, Q.; Xia, L.; Huang, D.; Fu, X.; Zhang, R.; Hu, S.; Zhao, F.; Li, X.; Bao, X. Poly(2,5-benzimidazole)/sulfonated sepiolite composite membranes with low phosphoric acid doping levels for PEMFC applications in a wide temperature range. *J. Membr. Sci.* **2019**, *574*, 282–298. [CrossRef]
83. Biron, M. Detailed Accounts of Thermoplastic Resins. *Thermoplast.Thermoplast. Compos.* **2018**, 203–766. [CrossRef]
84. Petreanu, I.; Marinoiu, A.; Sisu, C.; Varlam, M.; Fierascu, R.; Stanescu, P.; Teodorescu, M. Synthesis and testing of a composite membrane based on sulfonated polyphenylene oxide and silica compounds as proton exchange membrane for PEM fuel cells. *Mater. Res. Bull.* **2017**, *96*, 136–142. [CrossRef]
85. Liu, X.; Zhang, Y.; Chen, Y.; Li, C.; Dong, J.; Zhang, Q.; Wang, J.; Yang, Z.; Cheng, H. A superhydrophobic bromomethylated poly(phenylene oxide) as a multifunctional polymer filler in SPEEK membrane towards neat methanol operation of direct methanol fuel cells. *J. Membr. Sci.* **2017**, *544*, 58–67. [CrossRef]
86. Zhu, X.; Shen, C.; Gao, S.; Jin, H.; Cheng, X.; Gong, C. High-temperature proton exchange membrane with dual proton transfer channels by incorporating phosphonic acid functionalized siloxane into poly(2,6-dimethyl-1,4-phenyleneoxide) (PPO). *Solid State Ion.* **2019**, *337*, 193–204. [CrossRef]
87. Jang, J.; Kim, D.H.; Ahn, M.K.; Min, C.M.; Lee, S.B.; Byun, J.; Pak, C.; Lee, J.S. Phosphoric acid doped triazole-containing cross-linked polymer electrolytes with enhanced stability for high-temperature proton exchange membrane fuel cells. *J. Membr. Sci.* **2020**, *595*, 117508. [CrossRef]
88. Ahn, K.; Kim, M.; Kim, K.; Ju, H.; Oh, I.; Kim, J. Fabrication of low-methanol-permeability sulfonated poly(phenylene oxide) membranes with hollow glass microspheres for direct methanol fuel cells. *J. Power Sources* **2015**, *276*, 309–319. [CrossRef]

89. Wu, W.; Zou, G.; Fang, X.; Cong, C.; Zhou, Q. Effect of Methylimidazole Groups on the Performance of Poly(phenylene oxide) Based Membrane for High-Temperature Proton Exchange Membrane Fuel Cells. *Ind. Eng. Chem. Res.* **2017**, *56*, 10227–10234. [CrossRef]
90. Motealleh, B.; Huang, F.; Largier, T.D.; Khan, W.; Cornelius, C.J. Solution-blended sulfonated polyphenylene and branched poly(arylene ether sulfone): Synthesis, state of water, surface energy, proton transport, and fuel cell performance. *Polymer* **2019**, *160*, 148–161. [CrossRef]

MDPI
St. Alban-Anlage 66
4052 Basel
Switzerland
Tel. +41 61 683 77 34
Fax +41 61 302 89 18
www.mdpi.com

Polymers Editorial Office
E-mail: polymers@mdpi.com
www.mdpi.com/journal/polymers

www.ingramcontent.com/pod-product-compliance
Lightning Source LLC
LaVergne TN
LVHW070905170726
843515LV00005B/708

* 9 7 8 3 0 3 6 5 2 9 4 1 7 *